#222218

The Physiology of Flowering

Volume I
The Initiation of Flowers

Authors

Georges Bernier, Ph.D.
Professor of Plant Physiology
Director, Centre de Physiologie Végétale Appliquée (I.R.S.I.A.)
University of Liège
Liège, Belgium

Jean-Marie Kinet, Ph.D.
Assistant
Centre de Physiologie Végétale Appliquée (I.R.S.I.A.)
University of Liège
Liège, Belgium

Roy M. Sachs, Ph.D.
Professor and Plant Physiologist
Department of Environmental Horticulture
University of California
Davis, California

Ref QK 830 .B47 1981 vol. 1
Bernier, Georges, 1934-
The initiation of flowers

CRC Press, Inc.
Boca Raton, Florida

Library of Congress Cataloging in Publication Data

Bernier, Georges, 1934-
The initiation of flowers.

(The Physiology of flowering; v. 1—2)
Bibliography: p.
Includes index.
1. Plants, Flowering of. I. Kinet, Jean-Marie, joint author. II. Sachs, Roy M., joint author.
III. Title. IV. Series: Physiology of flowering; v. 1—2.
QK830.B47 582.13′04166 80-24351
ISBN 0-8493-5709-8 (v. 1)
ISBN 0-8493-5710-3 (v. 2)

Direct all inquiries to CRC Press, Inc., 2000 Corporate Blvd., N.W., Boca Raton, Florida 33431.

Second Printing, 1985
Third Printing, 1985

International Standard Book Number 0-8493-5709-8 (Volume I)
International Standard Book Number 0-8493-5710-3 (Volume II)
International Standard Book Number 0-8493-5711-X (Volume III)

Library of Congress Card Number 80-24351
Printed in the United States

OBJECTIVES AND ORGANIZATION OF THE WORK

Scientific advances often come from uncovering a hitherto unseen aspect of things as a result, not so much of using some new instrument, but rather of looking at objects from a different angle.

F. Jacob*

From the point of view of basic biology, plant scientists have given much attention to flowering because this is the first step towards sexual reproduction in plants. Also, since the onset of flowering is under absolute environmental control in many species, some biologists with a deep interest in morphogenesis chose the flowering process as their field of research. Interest in this developmental step has a strong economical basis, too, since many aspects of agronomic and horticultural crop production are intimately associated with flowering. Fundamental investigations provide a conceptual framework for the development of practical applications, particularly following the discoveries of photoperiodism and vernalization early in this century. From this period to the present, research has expanded rapidly. Just before World War II, the concept of the floral hormone or florigen was defined; this idea has received experimental support and is apparently the most widely accepted theory for the control of floral initiation. Despite numerous attempts to isolate and identify the hypothetical hormone, we still have no idea of its chemical nature.

With the introduction of more refined investigation techniques in the field, it is apparent that the flowering process is extraordinarily complex. It includes several interrelated steps, each of which is influenced by several factors of both internal and external origin. The simple florigen concept seems no longer commensurate with the complexity of the phenomenon it was supposed to explain. Possibly because of its simplicity it has remained the favorite theory. The physiology of flowering is thus in the uncomfortable situation, perhaps unique in biology, that it is still dominated by a concept proposed more than 40 years ago. It is recognized more and more that this concept puts severe constraints on further developments in both fundamental and applied research programs in which the onset of flowering is the key process.

In all experimental sciences, the appearance of new evidence that is irreconcilable with a theory, no matter how well established the theory may be, requires consideration of alternate hypotheses. Following this rule we suggest that a reconsideration of accepted notions of the physiology of flowering is absolutely and urgently required.

The present work is organized such that the whole phenomenon of flowering is divided into two major steps: (1) the initiation of flower primordia and (2) the development of these primordia into mature flowers until anthesis. Despite the fact that a separation of these two stages does not appear possible on theoretical or evolutionary grounds, the absence of clear distinction between the various stages of the flowering process in several investigations makes interpretation difficult or impossible, and therefore, the results are of little value. These stages are not all alike and do not always react similarly to external and internal variables. They should thus be considered independently to avoid confusion.

The first two volumes are devoted to "flower initiation" which includes not only the production by meristems of clearly recognizable flower primordia, but also all preceding reactions that are required if flowers are to be initiated. This has been by

* From Jacob, F., *Science,* 196, 1161—1166, 10 June 1977. Copyright 1977 by the American Association for the Advancement of Science.

far the most studied stage because many physiologists view it as the critical turning-point from vegetative to reproductive growth.

Volume I is concerned essentially with a review and critical analysis of the classical data and concepts. The aim here is to pinpoint the firmly established facts and controversial issues as well as to stress the shortcomings of classical work and interpretations. Research has indeed focused very heavily throughout the past 60 years on the effects of physical and chemical factors of the environment, while unfortunately little attention has been paid until quite recently towards gaining an understanding of the basic internal mechanisms that underlie the floral transition.

The first section of Volume II deals with these more intimate aspects of the onset of reproductive growth. The basic role of correlative influences in flower initiation, even in the simplest experimental systems, is demonstrated. Then, we proceed by a description of the floral transformation of shoot apices at levels ranging from the macromorphological to the molecular. The idea is that a rather complete description is central to an understanding of the process of flower initiation and that it may further give some insights into the controlling agents of this process.

The second section of Volume II begins with a search for exogenous chemicals that control in part, or totally, the events of the floral transition. This is followed by a review of the work on endogenous substances that are considered as possible promoters or inhibitors of flower initiation. An attempt is made to see how far we have come in the understanding of the ultimate processes whereby a meristem begins to initiate flowers instead of leaves. The not surprising conclusion is that we are still a long way from the goal, but despite the fragmentary nature of the available evidence, the analysis developed in Volume II may provide a useful conceptual framework for future investigations in this important area of plant science. Also, it is anticipated that this new approach will result in development of more rational and efficient controls of flowering for agricultural and horticultural purposes. These applied aspects will be discussed in the third volume of this series.

The treatment of the different chapters is neither simplistic nor exhaustive. Our general philosophy has been to avoid extreme positions, either abusive generalizations that mask the real complexity of the problem and the diversity of plant behaviors or complete descriptions of all possible types of plant responses that create confusion and discourage the readers. Evidently, when one attempts to cover such an extensive subject in a limited number of pages there is inevitably a problem of topic selection. While our aim was to provide a balanced account of the most important and recent contributions in all aspects of the subject, some topics have wittingly received special treatment. Their selection reflects essentially our personal interest; other writers would have certainly made other choices and presented a differently balanced book. We like to think, however, that it will be recognized as timely and essential for the field to be reexamined from widely divergent points of view.

It is important to underline that constant reference to source material and use of a rich illustration should assist the unspecialized reader to obtain a full understanding of the discussed topics. Concluding sections are also inserted in many places and hopefully will be considered as resting spots. The busy reader may begin with these sections and the short Chapter 9, Volume II, and return to the main text for examination of important details. A glossary is also included for the reader who is unfamiliar with the scientific jargon of the field. In a work like this, there is some unavoidable repetition of material, but this has been reduced by frequent use of cross references. The species most commonly used in flowering studies will be usually referred to by their generic names alone.

The third volume will be concerned with the stages of flowering that follow initiation, essentially flower organization and maturation until anthesis. While the necessity of considering separately these successive stages was stressed above, it is also obvious that flowering is a unitary phenomenon and that its component steps are necessarily related to one another. After all flowering is about sexual reproduction in plants and we must expect that in evolution the entire physiological process is designed to expedite recombination of genetic characters and reproduction of the organism.

In this work we deal mainly with angiosperms, although gymnosperms are occasionally considered.

We hope that these three volumes will convey some of the excitement that we have felt during their preparation as well as during our investigations on flowering.

THE AUTHORS

Georges Bernier, Ph.D., is Professor of Plant Physiology, University of Liège, and Director of the Centre de Physiologie Végétale Appliquée (I.R.S.I.A.), Liège, Belgium.

Dr. Bernier graduated in 1956 from the University of Liège, with a "license" degree in botany and obtained his doctoral degree in 1963 from the same university. He performed an in-depth investigation on the uses of plants in nutrition, medicine, and other aspects of life in semiprimitive human societies in Zaire (1957 to 1958). He worked in the laboratories of Professors Buvat and Nougarede, Ecole Normale Superieure, Paris, France (1960 to 1961), and of Professor Jensen, University of California, Berkeley (1964 to 1965).

Dr. Bernier is a member of the American Society of Plant Physiologists, French Society of Plant Physiology, Society for Experimental Biology (U.K.), Federation of European Societies of Plant Physiology, Belgian Society of Biochemistry, and the Belgian Society of Cell Biology.

Dr. Bernier has presented many invited lectures at International Meetings, Universities, or Institutes. He has published more than 70 research papers and book chapters, and a book on *Cellular and Molecular Aspects of Floral Induction* (Longman, London, 1970) including the proceedings of a symposium that he organized at the University of Liège. His current major research interests include the molecular and cellular changes occurring in meristems at floral evocation and subsequent flower initiation, and the nature of the factors that control these developmental steps.

Jean-Marie Kinet, Ph.D., is a member of the Centre de Physiologie Végétale Appliquée (I.R.S.I.A.), Liège, Belgium.

He graduated in 1964 from the University of Liège, with a "license" degree in botany and obtained his doctoral degree in 1972 from the same university. He worked in the laboratory of Professor Wareing, University College of Wales, Aberystwyth, U.K. (1975).

Dr. Kinet is a member of the American Society for Horticultural Science, Belgian Society of Cell Biology, Federation of European Societies of Plant Physiology, and the International Society for Horticultural Science.

He has published more than 30 research papers. His current major research interests include the physiological and cellular aspects of flower initiation, inflorescence and flower development, and the inference that fundamental studies have on research concerning species with an economical interest.

Roy M. Sachs, Ph.D., is Professor and Plant Physiologist, Department of Environmental Horticulture, University of California, Davis.

Dr. Sachs graduated in 1951 from the Massachusetts Institute of Technology, Cambridge, with a B.Sc. in Biology; obtained the Ph.D. degree in Plant Physiology in 1955 from the California Institute of Technology, Pasadena, under the guidance of Dr. Went. In 1958 he began his academic career with the University of California, Los Angeles (1958 to 1961) and was with the Department of Floriculture and Ornamental Horticulture, and moved to Davis to join the Department of Landscape Horticulture (later renamed Environmental Horticulture). He was a Fulbright Fellow at the Università di Parma, Instituto di Botanica, Italy in 1955 and 1956 working with Dr. Lona. From 1956 to 1958 he was a Research Botanist at the University of California, Los Angeles with Dr. Lang. In 1969 to 1970 he was a visiting research scientist at the Vegetation Control Laboratories, U.S. Department of Agriculture (USDA), Fort Detrick, Frederick, Md., in 1975 a North Atlantic Treaty Organization (NATO) visiting

Professor at the Facoltà di Agraria, Universita di Padova, Italy, and in 1979 a visiting Research Scientist at the Département de Botanique, University of Liège, Belgium.

Dr. Sachs is a member of the American Society of Plant Physiologists Association and editor of *Scientia Horticulturae.* He has been the recipient of numerous grants for research on control of vegetative and reproductive development in higher plants. He has presented over 20 invited lectures on the control of vegetative and reproductive development.

Dr. Sachs is the author of more than 100 publications in scientific and technical journals, including review articles on reproductive development in higher plants. His current major research interests are in the areas of reproductive development, biomass production, and fuel conversion systems.

ACKNOWLEDGMENTS*

In preparing this work we acknowledge the support of our colleagues, first in our own departments (M. Bodson, W. P. Hackett, A. Havelange, and A. Jacqmard), and secondly those who have contributed so much to the original data based on flowering. But in a particular way we have appreciated the efforts of H. A. Borthwick, M. Kh. Chailakhyan, P. Chouard, L. T. Evans, K. C. Hamner, W. S. Hillman, A. Lang, A. Nougarède, F. B. Salisbury, D. Vince-Prue, S. J. Wellensiek, and J. A. D. Zeevaart, who have "paid their dues" in writing provocative and sometimes encyclopedic reviews of the subject.

We are greatly indebted to A. Bouniols, A. Nougarède, and M. Tran Thanh Van for supplying unpublished materials, and to these and many copyright owners for permission to reproduce illustrations and data from their publications. These sources are fully acknowledged in the captions of the figures and tables. One of us (Georges Bernier) would especially like to express appreciation to the University of Liège for providing facilities during the preparation of this manuscript. The assistance of Michelyne Dejace and Maggy Grossard in typing the manuscript and preparing the graphs is also gratefully acknowledged.

* Quotation on divider page, Section I from Went, F. W., *Union Int. Sci. Biol. Ser. B,* 34, 169, 1957. With permission. Quotation on divider page, Section II from Wardlaw, C. W., *J. Linn. Soc. London Bot.,* 56, 154, 1959. With permission. Quotations on divider page, Section III from Lang, A., *Communication in Development,* Lang, A., Ed., Academic Press, New York, 1969, 244. With permission. From Letham, D. S., Higgins, T. J. V., Goodwin, P. B., and Jacobsen, J. V., *Phytohormones and Related Compounds: A Comprehensive Treatise,* Vol. 1, Letham, D. S., Goodwin, P. B., and Higgins, T. J. V., Eds., Elsevier/North-Holland Biomedical Press, Amsterdam, 1978, 1. With permission.

TABLE OF CONTENTS

Volume I

SECTION I — CLASSICAL EXPERIMENTAL SYSTEMS AND DATA; ANALYSIS OF COMMON CONCEPTS

TABLE OF CONTENTS

Volume II

SECTION I

Classical Experimental Systems and Data; Analysis of Common Concepts

> It is obvious that the problem of flowering [. . .] is a very complicated phenomenon, and [. . .] I do not believe that anything is gained by considering it simple.
>
> F. W. Went

Chapter 1

EXPERIMENTAL SYSTEMS

TABLE OF CONTENTS

I. PLANT MATERIALS

Flower initiation is best investigated in the minority of species or varieties in which this process is under relatively strict control of the environment, as in absolute photoperiodic and cold-requiring plants. Many investigations require that plants start to flower at the same time and are at the same stage of development. In these species during a period of growth in noninductive conditions, an apparent uniformity in response to inductive conditions is achieved giving the required synchrony.

In the majority of higher plants, photoperiod and temperature affect the rate of the transition to flowering, but do not control it in an absolute sense. In such species, the lack of strict developmental control usually hinders the design of critical experiments because there is no precise zero time and insufficient synchronization of plants that must be analyzed as a group. Facultative photoperiodic or cold-requiring plants, while more common, do not represent anything basically different from the absolute ones since many species may have absolute daylength or temperature requirements in one set of environmental conditions and exhibit facultative behavior when some of the parameters are changed.

Early work is mostly with photoperiodic plants requiring several inductive cycles, but recent investigators favor species requiring exposure to only a single photoinductive cycle, because they are presumably the species with the greatest achievable uniformity and ease of timing of postinductive events.

An often-raised question concerns the real significance of the observed structural and chemical changes in relation to the transition to flower formation: are these changes an integral part of the developmental switch or simply accompanying unrelated changes? An unambiguous answer usually requires considerable experimental work and even then is quite difficult to reach. Very often the way to simplify this problem is to examine comparatively developmental transition in a variety of physical or chemical environments and see whether the changes in question are always present. Fortunately in many species, including most of the "single-cycle" ones, flowers initiate in response to more than one experimental treatment (Table 1), and this property makes these systems more exploitable.

Depending on the kind of experiments to be performed, plants may have to meet other prerequisites besides responsiveness to "one-shot" photoinduction. Speaking generally, small size at the responsive stage is essential in experiments that involve many treatments and in studies using a spectrograph or any similar equipment where the area irradiated with any wavelength of light is very limited. Consequently, the SDP *Pharbitis nil*,[1] *Chenopodium rubrum*,[2] and the LDP *Brassica campestris*,[3] which can be induced to flower by a single appropriate photocycle only a few days after sowing are particularly well-suited for such experiments.

Small fast-growing duckweeds, such as the LDP *Lemna gibba*, strain G3, first used by Kandeler,[4] or the SDP, *Lemna paucicostata*, strain 6746 (formerly designated as *L. perpusilla* 6746)[5] offer quite the same advantages as these seedlings. For the busy physiologist, duckweeds have the additional and inestimable advantage that if the flowering index of the cultures cannot be evaluated immediately at the end of an experiment, they may be stored until examination in a refrigerator for periods up to 2 weeks without any change of this index.[5] In addition, duckweeds floating aseptically in test tubes on defined nutrient media present themselves as ideal material for determining the effects of nutrition, and more generally of the chemical environment, on the flowering process.

When translocation of promoters or inhibitors of flower initiation is to be followed, seedlings or duckweeds are, however, frequently not suitable because they lack a transporting system of sizable length. The SDP *Xanthium strumarium*,[6] LDP *Lolium te-*

Table 1
ALTERNATE PATHWAYS TO FLOWER INITIATION IN SOME SINGLE-CYCLE SPECIES

Response Type	Species	Pathway	Discussed in chapter
SDP	*Pharbitis nil*	Continuous light and poor nutrition	2
		Continuous light at low temperature	3
		Continuous light at low light flux	3
		Continuous light at high light flux	3
		Continuous darkness	3
	Lemna paucicostata 6746	LD + copper	2
		LD at low light flux	3
		Skeleton photoperiods	4
		LD + salicylic acid	13[a]
LDP	*Lolium temulentum*	SD and anaerobiosis during one night	3
		SD + GA_3	13[a]
	Sinapis alba	SD at high photon flux	3
		Continuous darkness	3
		SD at low temperature	3
		Displaced SD	4

[a] Volume II, Chapter 6.

mulentum,[7] and *Sinapis alba*,[8] which respond to a single inductive cycle are more appropriate for such studies. These plants are also suitable for investigating changes associated with increased sensitivity to induction with age.

An important advantage, alluded to earlier, in using plants grown in strictly noninductive regimes is that such plants may be in a *more or less steady state of growth* during the experimental period. They usually produce leaves of the same size and shape and at a constant rate, their stem elongates at near constant rate, and their meristem may be at a steady state with respect to size and growth rate, as in *Perilla*,[9] *Chenopodium amaranticolor*,[10] *Xanthium*,[11] *Silene coeli-rosa*,[12] and *Sinapis*.[13] On the other hand, seedlings during germination, do not lend themselves to certain kinds of postinductive analyses precisely because these growing characteristics are changing quite rapidly in both the induced and noninduced plants.

Studies on floral evocation and morphogenesis are more significant with species producing a terminal flower or inflorescence since the complete transformation can be traced in these plants in one and the same meristem. Plants having a decussate or distichous arrangement of leaves at the vegetative stage are especially appropriate for microscopic work that requires an accurate orientation of the meristem for sectioning.

Species requiring a single photoinductive cycle, interesting as they are, may not be the best material for gaining insight into some particular problems. For instance, the processes of fractional induction and partial evocation are easiest to attack in species requiring several inductive cycles or having a dual photoperiodic requirement as LSDP and SLDP.

Certain studies are possible only when the plant material possesses still additional attributes, for example, ease of grafting to a partner. Unfortunately, this attribute is not common in "single-cycle" plants, with the exception of *Xanthium*.

Quite often unsatisfactory results are obtained because of genetic heterogeneity of the plant material, the seed having been commercially purchased or collected from wild plants. In these cases a genetically uniform material must be produced by clonal propagation or inbreeding before the work itself can start. Interest in using plants in which the genetics of flowering are well-known is self-evident, but unfortunately there are only a few species for which this knowledge is available, and these are not "single-cycle" plants.[14]

CONCLUSIONS

Clearly, the *ideal* warhorse for the study of flower initiation does not exist since requirements for one kind of experiment are opposite to those for another. However, particular materials are certainly far more appropriate than others to investigate in detail a specific problem. The choice by an investigator of the plant(s) most appropriate for his work can only be made after a careful analysis of all the facets of his research project. The importance of this step, preliminary to the experimental work itself, should not be underestimated.

Yet, some workers may find it more attractive to look for new experimental materials. The number of species in which flowering has been the subject of research is still quite small and new remarkable plants, perhaps more advantageous for laboratory work than the presently favored experimental systems, remain to be discovered. Also, geneticists may introduce new genotypes well-suited to particular physiological purposes. However, an understanding of the basic mechanisms of flower initiation will only arise from an in-depth study of one or a very small number of experimental systems. It is clearly apparent from other biological branches, e.g., molecular biology of prokaryotic organisms or genetics, that major advances were obtained by a cooperative intensive work on a single selected biological object, such as *Escherichia coli* or *Drosophila melanogaster.* Thus, although it is acknowledged that the search or production of new experimental plants is a very valuable task that has to be pursued, it is hoped that physiologists will realize that they already have in hand many excellent systems and that a majority of the incoming workers will investigate these systems.

II. MEASUREMENT

An experimental approach to any biological process evidently requires its accurate measurement, and progress is often dependent on refinements of the measurement technique. Work on flowering is in no way an exception to these generalizations. However, despite manifold attempts to achieve a universally accepted system, the measurement of flowering remains as varied as the species under investigation. Flowering measurements are always based on morphological changes at the apical meristem which are the results of several earlier processes occurring in different plant parts. Accordingly, measurements do not permit discrimination among these various steps.

Also, several characters of the flowering response that can be used to measure this process are affected by environmental conditions prevailing during the life of the plants. For instance, mineral nutrition is known to affect markedly the number of flowers in a variety of species (see Volume I, Chapter 2, Table 1). Clearly then with the progressive refinements of the experiments and measurement techniques, there is an increasing need to control more and more carefully the environmental parameters, that is to grow the plants during their *whole* life cycle in growth chambers or phytotrons, possibly on nutrient solutions and in a controlled atmosphere. However, because space under rigid environmental control is usually restricted, the plants are often grown in a greenhouse (with a partial control of temperature and light) before the period during which long- or short-term treatments are given. This is frequently the case when a large number of experiments or a large number of replicate plants are needed. However, experiments of this kind with many species have shown that previous treatment affects subsequent performance of the plants, and the results are clearly season-dependent. While these difficulties of cultivation cannot always be avoided, they reduce very much the reproducibility of the results and prevent absolute comparisons between experiments made at different times. They also put severe limitations to the precision that can be achieved in the work.

A. Flower Initiation: A Qualitative or a Quantitative Phenomenon?

Consider that (1) a plant is either flowering or not and (2) flowering is more abundant in a plant producing ten flowers than in the one producing only one flower. Thus, there is both a qualitative and a quantitative component.

The classical view is that the flower is nothing other than a compressed, determinate shoot bearing sepals, petals, stamens, and carpels as lateral organs instead of leaves. This concept of homology between vegetative and reproductive shoots goes back to the ideas of Goethe in 1790 and is still widely accepted by morphologists of today. Indeed, in certain groups of angiosperms considered to be primitive, the growth of the floral meristem is indeterminate, and a large number of floral parts is produced. These floral organs occur on a rather elongate axis and are arranged helically. Transitional forms between leaves and bracts, bracts and perianth parts, sepals and petals, and petals and stamens are not uncommon in the normal morphology of many plants as well as in numerous teratological forms.

Some anatomists believe that reproductive and vegetative shoot meristems are structurally and functionally alike implying that there is no basic difference between their immediate derivatives. If accepted as such the concept that a flower is a short shoot bearing modified leaves casts doubt about the qualitative nature of the flowering process. Physiologists, seeing no discontinuity between the stimulatory effect of minimal and optimal induction, have also argued that floral initiation is a quantitative response only.[15]

However, the problem is to determine if the differences, not only the structural differences but also the functional and chemical ones, between a foliage leaf and a floral part are sufficiently large to consider these two organs as qualitatively different entities. Several studies have demonstrated that the protein complement of flowers and flower parts is qualitatively different from that found in vegetative organs of the same species, and there are metabolic pathways peculiar to flower organs, e.g., those leading to synthesis of the pollen wall components and pigments of perianth parts. Thus, considering functional aspects, one concludes that a flower is composed of parts that are, without any doubt, qualitatively different from true leaves. Most probably, as will be discussed in Volume II, Chapter 4 genes that are inactive in vegetative organs are activated in flower organs. Refined studies concerning shoot apical meristems show that reproductive and vegetative meristems, even when they look superficially alike, exhibit marked functional, ultrastructural and chemical differences (Volume II, Chapters 3 and 4). These differences are clearly detectable before the onset of flower initiation. Viewed that way, the process of flower initiation has basically a qualitative nature. We know now that in subminimal inductive conditions some changes occur in the plant (fractional induction, partial evocation), but they almost never result in the production of a fraction of a flower. As a rule, flower initiation starts when evocation has reached a certain threshold and then proceeds until the formation of at least one or a few morphologically distinct floral structures; this process has thus an *all-or-none* aspect. Beyond the first flower the intensity of flowering may vary greatly depending on many factors, and the process of flower initiation has thus also a *quantitative* aspect.

B. Advantages and Limitations of Various Measurement Methods

The simplest qualitative measure of flower initiation is the proportion of plants that have initiated at least one flower after a given treatment. For the individual plant, the scoring is thus either + or −, and for the experimental population, the index of flower

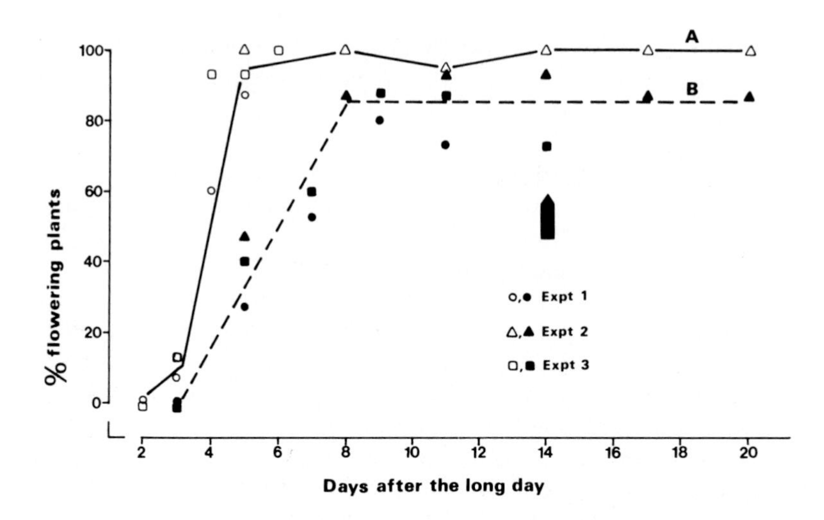

FIGURE 1. Percent of plants having initiated flower primordia in *Sinapis alba* as a function of time after the start of an inductive 20-hr (open symbols) or a 14-hr (dark symbols) LD. Routinely, the plants are dissected, and their apical meristem examined 14 days after the start of the experimental treatment (arrow). Notice that at this time the index of flower initiation has reached an unchanged value.

initiation is given by the percent of + individuals. Initially, investigators waited until macroscopic appearance of flower buds or even opening of flowers before determining this index. Since conditions favoring flower growth and development are not always the same as those producing initiation and since flower primordia might not develop beyond a certain stage, the index of flower initiation is now usually determined by dissection under a binocular microscope at an appropriate time after the experimental treatment (Figure 1). Measurements of this kind are usual in species like *Pharbitis*, *Chenopodium*, *Sinapis*, *Anagallis*, and duckweeds. Timing of the examination is especially critical in duckweeds in which flower buds may eventually abort and finally disappear in conditions of minimal induction.[16] A disadvantage of this dissecting method is that it requires the destruction of the examined plants.

While often entirely satisfactory, this method depends in fact upon there being some residual genetic and/or environmental variation among the test plants. Were they wholly uniform in behavior, the population scores would always be close to 100% floral or 100% nonfloral. In such homogeneous materials, like the violet strain of *Pharbitis*, Biloxi soybean, and the clonal populations of *Anagallis*, the number of initiated flower buds (or flowering nodes) provides then the quantitative scale required. This index offers a further advantage in that a value is obtained for each plant within a batch.

There are, however, a number of limitations in using flower number. In some species, this number is not modified by treatments because it is primarily influenced by the structure of the reproductive axis, an inherited trait. Second, in plants producing, for example, an indefinite raceme, like *Sinapis*, or a capitulum, like *Xanthium* and *Chrysanthemum*, a large number of small flower buds is generally produced even after a minimal induction, and a count of these buds is difficult and tedious.

In other studies, time between start of the experimental treatment and attainment of a certain flowering stage, e.g., the initiation of flower primordia, macroscopic appearance of flower buds, or anthesis is often recorded. When using flower appearance or anthesis, it must be realized however that the character actually measured is composed of several components, including among others the rate of flower development.

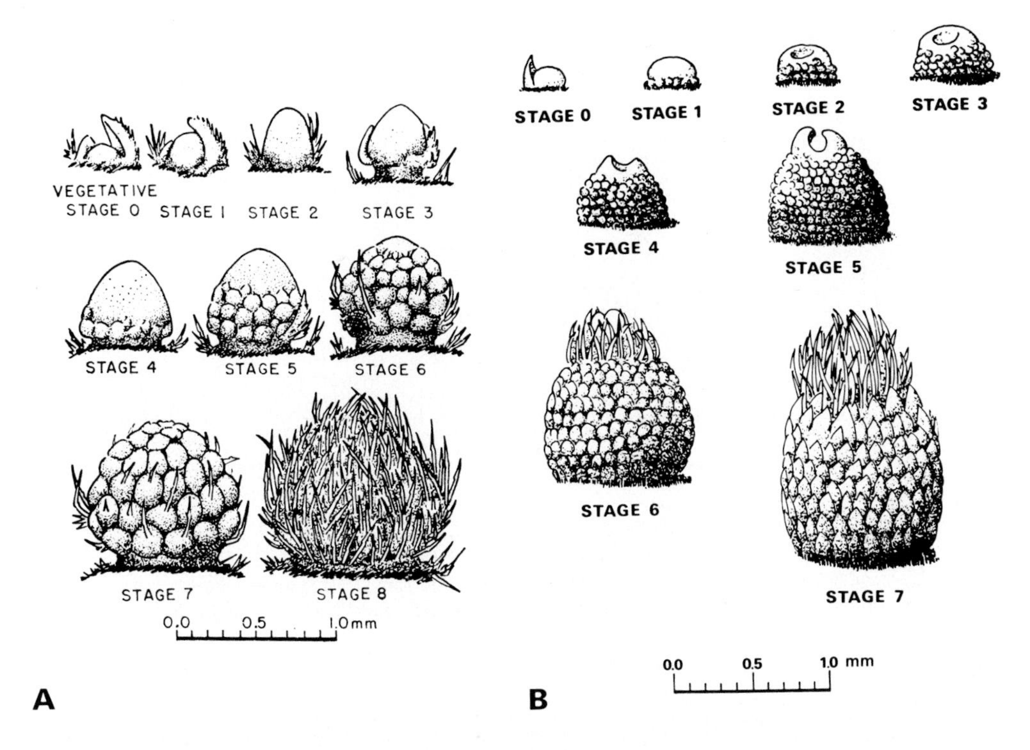

FIGURE 2. Floral stages in *Xanthium strumarium.* (A) For the staminate inflorescence. (From Salisbury F. B., *Plant Physiol.,* 30, 327, 1955. With permission.) (B) For the pistillate inflorescence. (From Léonard, M., Kinet, J. M., Bodson, M., Havelange, A., Jacqmard, A., and Bernier, G., *Plant Physiol.,* in press.)

A basic postulate is that this rate is proportional to the degree of floral induction and that both initiation and development are influenced in the same way by the inductive treatment. These assumptions are certainly valid in many cases, but as indicated earlier, they do not apply to all species. When the experimental treatment begins early in the life of the plant, at seeding for instance, days to bloom also include the rate of plant passage through the juvenile phase.

''Floral stage'' systems in which successive microscopic and sometimes macroscopic stages of flower or inflorescence development are arbitrarily assigned increasing numerical values are widely used. At some arbitrary time after the start of the experiment, the meristem of each plant within an experimental group is examined and, based upon its developmental stage, is assigned a number. The resulting values for the group are then averaged. Although partly subjective and usually destructive because it requires dissection of plants, this kind of measuring system is very sensitive and reduces considerably the period between the treatment and the examination of plants.

The most famous of these scoring systems is that designed by Salisbury for the terminal staminate inflorescence of *Xanthium* (Figure 2A).[17] A similar system has been recently elaborated for the pistillate lateral inflorescences of the same species (Figure 2B).[18] Other similar systems have been devised for many species, including *Chrysanthemum, Silene, Anagallis,* and *Bryophyllum.*

Salisbury[19] has long insisted that a very desirable property of these systems is that they produce nearly straight lines on a graph when the floral stages are plotted as a function of time after the start of induction or when floral stages at some arbitrary time following induction are plotted as a function of the effectiveness of induction (Figure 3).

However, useful and successful they may be, these systems are nothing but a measurement of the rate of inflorescence development as clearly seen in Figure 2. In the

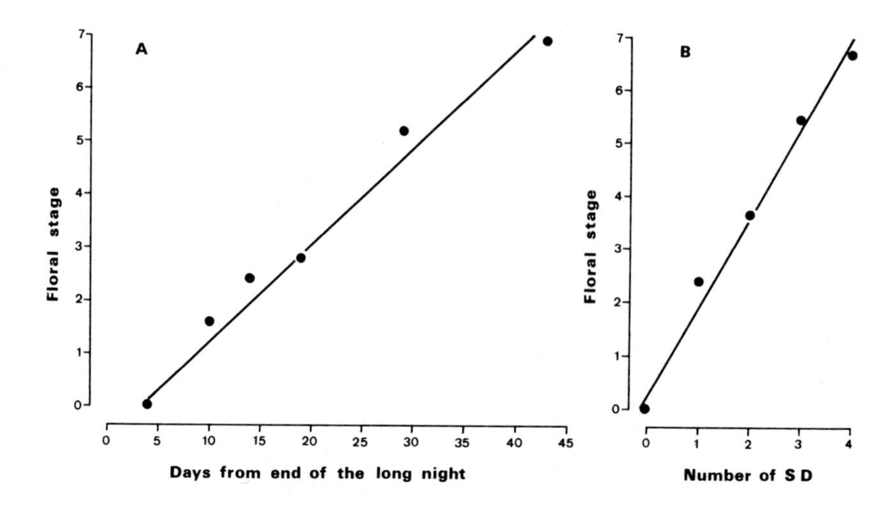

FIGURE 3. Axillary female flowering at node 8 (numbered acroetally from the induced leaf) in *Xanthium strumarium*. (A) As a function of time after end of the single inductive long night; (B) as a function of strength of induction (number of inductive long nights given to the plants). Results in B are collected 2 weeks after end of the first long night.

long run the possibility remains thus that they may lead to some confusion, and we must be aware of this problem.

An additional difficulty may arise with systems which include some vegetative stages. For example, stages 1 and 2 of Salisbury's system for *Xanthium* (Figure 2A) are certainly vegetative since they are known to revert to Stage 0 (which is typical of a strictly vegetative meristem) under noninductive conditions.[17] Thus, in cases of marginal induction in *Xanthium,* the individual floral stages in a group of plants may be relatively similar when examined 10 days after the treatment, but can be totally dissimilar when examination is made 1 month later: part of the population has then reverted to stage 0 whereas the other plants have reached stage 7 or 8. Moreover, even though stages 1 and 2 are normal intermediates in the floral transition of the meristem, they can also be observed in plants which never flower, e.g., after a GA_3 treatment in noninductive conditions.[11]

In several Gramineae, e.g., rye and *Lolium,* and in *Chenopodium,* the length of the shoot apex, which can be measured very accurately, is often used as an index of flower initiation. In *Lolium,* it has been found indeed that the initiated inflorescence increases in length exponentially with time and as function of the intensity of induction, e.g., the number of inductive cycles given to the plants (Figure 4).[20] Although the elongation (mounding up) of the meristem is one of the changes which is most typical of evocation, its use as an assay of flower initiation can be misleading since it may occur in the absence of flowering in *Chenopodium,*[2] as well as in many other species (Volume II, Chapter 2). On the other hand, Cumming has reported that some treatments, like sugar application, that are fully inductive in *Chenopodium* may partially inhibit the increase in size of the meristem.[2]

Because of these limitations, the length of the shoot apex should always be used in conjunction with another measure of the initiation of flowering.[7]

C. Measurement Based on Previous Vegetative Growth

Using experimental systems which require a protracted period of induction or using DNP, the rate of development is often measured by the number of nodes (leaves) preceding the first flower (or inflorescence). This measure of flower initiation was first used in cereals and then in many other plants, including subterranean clover, pea,

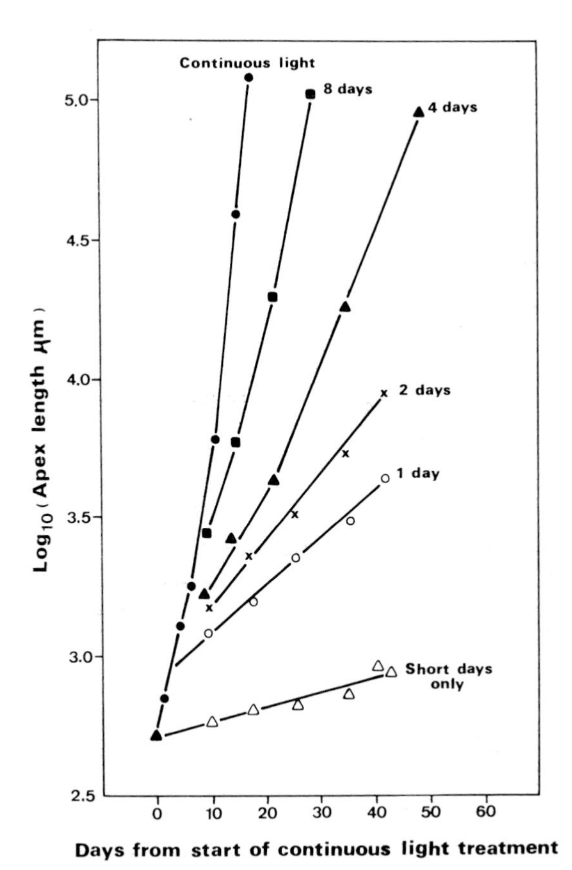

FIGURE 4. Inflorescence development, as estimated by apex length in the LDP *Lolium temulentum* receiving continuous illumination for the various periods indicated. (From Evans, L. T., *New Phytol.*, 59, 163, 1960. With permission.)

soybean, strawberry, chrysanthemum, tomato, and hemp. When the plant possesses a large number of nodes (leaves) at the start of the experiment, instead of counting the absolute node number for the first flower, it is often preferable to determine the difference between this and the youngest node present at the start of the treatment, that is "node increment".[21] Such a widespread use comes from the fact that a very precise and totally objective value is obtained for each individual plant and because it was believed that the results expressed in this manner are independent of the effects of the treatments on the growth rate. Indeed, the node of first flowering and the time of flower initiation may be correlated in many cases. However they do not always keep strictly in step with each other because conditions which modify profoundly the rate of node initiation, that is the rate of vegetative growth, do not necessarily affect the time of flower initiation. Thus, there are instances of reduction in node number without any advance in flower initiation and of increase in node number without any delay in flower initiation.[14,22] For node number not to be a misleading index of flower initiation, information on the rate of node production is also necessary in these cases.

A measurement of this kind is useless with most "single-cycle" plants since their apical meristem usually stops vegetative growth very rapidly after induction.

In some cases leaf numbers on axillary shoots may be more sensitive to treatments than leaf number on the main shoot.[23]

CONCLUSIONS

A variety of methods have been used to measure flower initiation. Ideally each method should evaluate both the qualitative and quantitative characters of this developmental process, but this goal is rarely achieved. Therefore, whenever possible, more than one method should be used with one species.

The most appropriate measuring techniques are chosen with respect to the species investigated, the aim and precision of the work, the number of plants available, etc. The selected methods, whatever they are, are not without potential error. Several techniques, including the "floral stage" systems, are in fact based on the measurement of the rate of flower or inflorescence growth and development. In order to reduce the possibility of interference by effects on processes subsequent to initiation, microscopic examination of the plants should be made at the shortest appropriate time interval after the treatment.

The determination of the first flowering node, a method which is sometimes used to avoid interference of the growth rate, also has limitations. Another danger is encountered with methods measuring a morphological feature, such as meristem swelling, which is usually an integral part of the process of flower initiation but which may also occur in the complete absence of flower production.

Since most measuring techniques are based on formation of flower buds, it must be emphasized that we have no independent assay of leaf induction.

Chapter 2

CONTROL BY NUTRITION AND WATER STRESS

TABLE OF CONTENTS

I. THE CARBOHYDRATE/NITROGEN RELATIONSHIP THEORY

Early in this century, before the discovery of photoperiodism and vernalization, ideas about the flowering process were much influenced by observations and practices made in the horticultural and agricultural fields. In several crops, fruit trees for instance, luxuriant vegetative growth is usually antagonistic to flowering. Treatments that reduce this vegetative growth, e.g., drought, pruning, girdling, often markedly promote flowering. Likewise, a high mineral supply, especially that of nitrogen, may reduce reproductive development in some plants, and incidentally enhance vegetative growth. On the other hand, Klebs and others observed that in many plants conditions favoring flowering were always favorable also for photosynthetic carbon dioxide fixation and thus led to a high carbohydrate level in the leaves.[24,25]

From these data, Klebs in 1913 proposed that flowering is controlled by the nutritional status of the plant, i.e., the balance of materials that the plant obtains from the air and the soil. A high endogenous ratio of carbohydrates to nitrogen (the C/N ratio) was believed to be essential for flowering.

Most of the data were collected in these early studies from plants in quite advanced stages of reproductive development and thus largely concern flower development and even fruit set rather than flower initiation. Another difficulty is that this general theory of the control of flowering evolved from experiments on plants, e.g., tomato, in which flower initiation is not subject to close environmental control.

Further studies with plants in which flower initiation is under the absolute control of the environment have generally failed to confirm the Kleb's theory. For example, in the SDP, Biloxi soybean, a high C/N ratio in the shoot is not evident until the pods are maturing.[25] Floral induction and inception of flower primordia in this plant occur when the ratio is *lower* than in vegetative plants.

Similarly, attempts to alter the C/N ratio in photoperiodic and cold-requiring plants by manipulating the N supply to the roots show that: (1) in many cases the basic inductive requirements are not altered by the N level. This is best illustrated in experiments of Naylor with *Xanthium* indicating that plants low or high in N have the same critical daylength for flower initiation,[26] and (2) the reaction of plants towards N is specific (Table 1), and thus that this element cannot play such a general deleterious effect as postulated by Klebs. As shown by Chailakhyan and others, in photoperiodic conditions that promote flowering, there are clearly plants in which flower initiation is favored by low levels of N, e.g., the LDP *Sinapis* and the SDP *Chenopodium rubrum,* and plants in which the same process is favored by high N levels, e.g., the LDP *Anethum graveolens* (dill), the SDP *Kalanchoe,* and *Perilla.*[2,27,28] Other plants, including the LDP *Spinacia oleracea* (spinach), the SDP soybean, and the DNP *Fagopyrum sativum* (buckwheat) are relatively insensitive to N supply. Flower formation in unvernalized plants of winter rye, but *not* in vernalized ones, is promoted by high N levels,[29] and this raises the possibility that mineral nutrition would be more important in cold-requiring plants, and perhaps also in photoperiodic species, in conditions that are only marginally inductive.

Other early observations on photoperiodic plants tended also to indicate that the initiation of flowers stands in no direct relationship to photosynthetic activity. Indeed, SDP are induced to flower by a reduction of the daily period of photosynthetic activity and in many LDP flower initiation is produced when the SD given at high intensity light are extended by a period of weak illumination which can hardly permit much photosynthesis.

On this basis, the importance of nutrition to flower initiation was soon discarded. Although the Kleb's theory appears now far too simple to account for the complexity of the flowering process, it had the great value of emphasizing the importance of both

Table 1
INFLUENCE OF N SUPPLY ON FLOWER INITIATION IN PHOTOPERIODIC PLANTS GROWN IN INDUCTIVE DAYLENGTHS

Species	Supply of N: Twice normal	Normal	Half normal	Null
	Days to visible flower buds			
Sinapis (LDP)	—	25	13	8
Anethum (LDP)	17	17	—	37
Spinacia (LDP)	10	10	—	14
Perilla (SDP)	23	23	24	32
Kalanchoe (SDP)	17	17	—	32
	Number of flowers			
Sinapis (LDP)	—	55	35	10
Perilla (SDP)	250	360	283	23
Kalanchoe (SDP)	210	217	—	12

Adapted from El Hinnawy, E. I., *Meded. Landbouwhogesch. Wageningen,* 56(9), 1, 1956.

mineral nutrition and photosynthesis in this process. Rejection of this theory had the unfortunate consequence that the role of nutrition was considered by many subsequent investigators as essentially trivial, restricted to the supply of raw materials, energy, and structural carbon skeletons needed for all processes occurring within the plant. However, results have accumulated over the years showing that the relationship of mineral nutrition and photosynthesis to flower initiation is not always so indirect and that, in some plants at least, it may even be regulatory. There are also recent indications of complex interactions between photosynthesis and mineral nutrition. Evidence concerning the role of photosynthesis and assimilate supply will be considered in Volume II, Chapter 8.

II. MINERAL NUTRITION: NITROGEN

Besides its modulating effect on flower initiation in plants grown in environmental conditions that promote flowering, mineral nutrition alters *qualitatively* the basic inductive requirements in a few species. In the cold-requiring plants, *Dactylis glomerata* and *Geum urbanum,* a high nutritional level, together with a high irradiance in the case of *Geum,* may substitute for vernalization.[30,31] In some photoperiodic plants, the critical daylength may be altered by changing the mineral level in the substrate. By decreasing the N level, in the presence of an optimal sucrose level in the culture medium, Deltour succeeded in obtaining 100% flower initiation in excised apices of the LDP *Sinapis* in SD.[32] In these conditions, apices flower after having produced the same number of leaves as they would have in LD. The SDP *Pharbitis* initiates flower buds under continuous light when the nutritional conditions are poor, while plants grown in the same conditions, but fed with nutrients do not.[33] The case of *Nicotiana glutinosa* is still more curious: this plant behaves as a quantitative LDP at a high level of nutrition and as a quantitative SDP at a very poor level. At an intermediate nutritional level it is a DNP.[34] Admittedly, few such examples are known, and data concerning vegetative growth in these cases are sometimes lacking. In interpreting experiments on nutritional influences, it is important to have such data in order to be sure that the basic requirements for growth are or are not met. The few examples cited

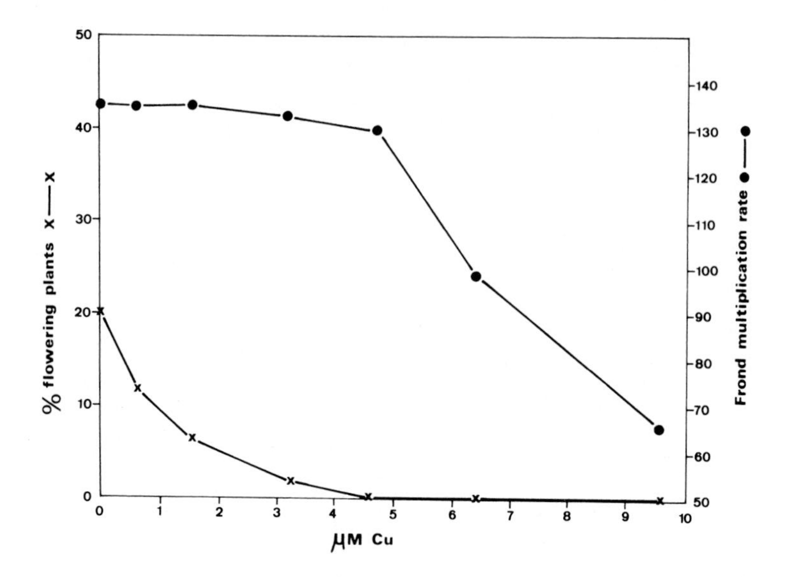

FIGURE 1. Effect of increasing copper concentration on the percent of flowering plants and the rate of frond multiplication in the LDP *Lemna gibba* G3 cultured in continuous light on a medium supplemented with 0.5 μg/mℓ of EDDHA. Control (0 concentration) contains 0.32 μmol/ℓ copper. (From Pieterse, A. H., Bhalla, P. R., and Sabharwal, P. S., *Plant Cell Physiol.*, 11, 879, 1970. With permission.)

above clearly suggest, however, that flower initiation, at least in some species, is rather directly tied to the availability of certain mineral elements. Nitrogen seems to be the active element in most of these cases.

III. MINERAL NUTRITION: TRACE ELEMENTS

Studies by Hillman with chelating agents proved that cupric ion is involved in the photoperiodic sensitivity of *Lemna paucicostata* 6746 and *L. gibba* G3.[35,36] *L. paucicostata* behaves as an absolute SDP when grown in Hutner's medium, which contains a chelating agent, and as a DNP in Hoagland-type medium *without* any chelating agent. The SD response is restored either by using the Hoagland's medium prepared in such a way that even traces of cupric ion are eliminated or by adding the chelating agent EDTA to the nonpurified medium. Supplementing the purified medium with low amounts of cupric ion eliminates the requirement for SD. Among other metal ions, Hg is less effective than Cu while Cd, Co, Cr, Mn, Ni, Pb, and Zn are ineffective.

Apparently parallel, but opposite, effects have been obtained with *L. gibba* which is unable to flower under any daylength in Hoagland's medium, but reacts as an absolute LDP with EDTA or when cupric ion is greatly reduced. Addition of Cu^{2+} to the purified medium inhibits flowering in LD.

Removal of excess Cu^{2+} in th medium, either by extensive purification or by addition of a chelating agent, potentiates the response to LD of both *Lemnas.* Thus, the chelating agents are seen as interfering in photoperiodic sensitivity through copper sequestration.

The promotive action of chelating agents, particularly EDDHA, on the flowering of *L. gibba* grown in LD was confirmed by Pieterse and co-workers who further showed that EDDHA is effective in permitting flower initiation of this LDP even in SD of 9 hr.[37,38] Again, Cu^{2+} is the metal involved in this effect since an increase in copper supply inhibits flowering totally in continuous light (Figure 1). This is further

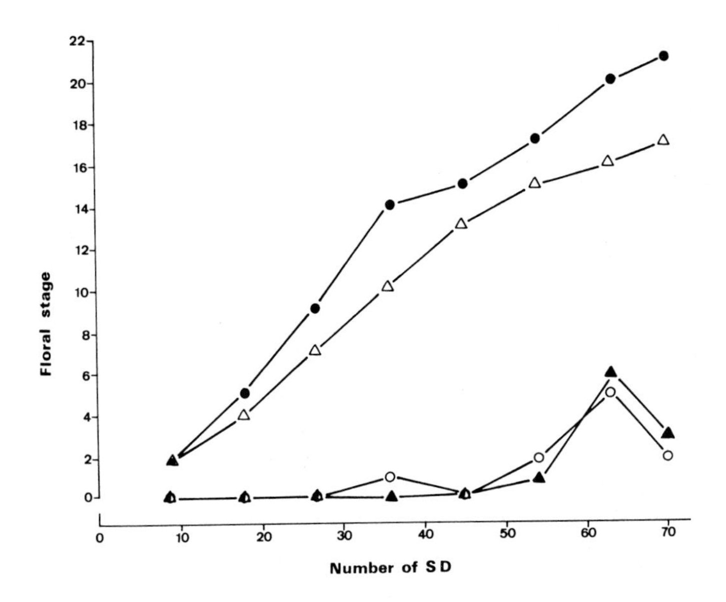

FIGURE 2. Effect of copper concentration in the nutrient solution on flower initiation and development on the second lateral shoot from the stem apex in *Chrysanthemum morifolium* cv. Pollyanne. ●-●: 0.05 ppm Cu; △-△: 0.001 ppm Cu; ▲-▲: 0.0005 ppm Cu; and O-O: no Cu. Flower initiation and development are measured by a system of floral stages. Stages 0 to 4: preparatory steps to initiation of floret primordia; Stages 4 to 8: initiation of floret primordia; and Stages 9 to 22: development of florets. (From Graves, C. J. and Sutcliffe, J. F., *Ann. Bot. (London)*, 38, 729, 1974. With permission.)

substantiated by the finding that the endogenous copper content is significantly lower in flowering plants of *Wollfia microscopica*, another duckweed, than in vegetative ones.[39] Excess copper may thus have a general perturbing effect on floral induction in duckweeds at concentrations that do not reduce frond multiplication (Figure 1).

The role of this element was also studied in several cultivars of *Chrysanthemum morifolium*.[40] Again copper was found to play an essential role in flower initiation. However, in this case, there is a direct relationship between the concentration of this metal in the nutrient solution and flower initiation. Below a level of about 0.0015 ppm initiation is almost completely prevented (Figure 2).

As noted, copper is probably involved in photoperiodic induction rather than in later stages of flower initiation in duckweeds. Hillman hypothesized that Cu interferes in some way with phytochrome action,[36] possibly affecting some metal-sensitive membrane system(s). Since: (1) copper is a SH-inhibitor, (2) other SH-inhibitors, such as mercury, silver, tungstate, and iodoacetamide, all promote flowering of *L. paucicostata* 6746 in LD and inhibit it in *L. gibba* G3,[41,42] and (3) the effect of all these compounds in *L. paucicostata* is nullified by cysteine,[43] Takimoto and Tanaka have for many years worked on the idea that some SH-enzymes are involved in photoperiodic sensitivity. On the other hand, copper might also regulate IAA level within the plant through its cofactor action on phenol oxydase activity;[40] IAA level may in turn control flower initiation (Volume II, Chapter 7). An understanding of the role of copper requires considerably more testing of these hypotheses.

Iron, too, seems critically involved in photoperiodic induction, at least in certain plants. Flower initiation in *Lemna paucicostata* 6746 grown in Hutner's medium in SD is strongly inhibited by reducing the iron supply while frond multiplication, i.e., vegetative growth, is little affected.[44] Thus, as pointed out by Hillman, iron is apparently essential for photoperiodic induction and is not regulating the flowering process simply through its more usually acknowledged role in general metabolism.

In *Wollfia microscopica,* rapid frond multiplication and flower initiation require the presence of a chelating agent in the culture medium. When the medium contains EDTA and ferric citrate, *Wollfia* responds as a typical SDP.[39] However, when these two compounds are replaced by Fe-EDDHA (an iron salt of EDDHA), vegetative growth remains vigorous, but the plant flowers now in LD as well as in SD. This effect of Fe-EDDHA is seemingly related to iron uptake by the fronds since plants grown in the presence of this chemical contain up to 50% more iron than those grown with EDTA and Fe-citrate.

Iron deficiency also prevents or greatly perturbs flower initiation in *Xanthium* while other elements are less essential.[45,46] Since flowering capacity is restored if iron is supplied previous to photoperiodic induction but not if given immediately after, this element is apparently needed during induction but not in further steps. The way iron interferes with induction is not known.

Other trace elements are apparently less essential than iron and copper, but exceptions to this rule are known. Molybdenum deficiency promotes flower formation in the SDP, *Lemna paucicostata* 6746 in LD,[47] this effect and that of copper addition described above being synergistic. Obviously, the relations between duckweed flowering and micronutrients are very complex, and the exact significance of the experimental results remains to be determined.

IV. WATER STRESS

The literature is rich in preliminary observations suggesting that water stress might be of importance for flower initiation in some plants. In general, however, experimentation has not proceeded far enough to reach definitive conclusions.

A remarkable exception is the experimental work of Bronchart on *Geophila renaris,* a perennial herbaceous plant of the tropical rain forest.[48] When grown from seeds in conditions of almost constant daylength and temperature (these are the normal conditions in the habitat of *Geophila*), this plant never flowers if the water supply is kept at field capacity. A period of water shortage of sufficient length, corresponding to a decrease of 30 to 50% of the available water, is absolutely required for flower initiation. The floral response is localized at certain axillary bud sites, and the proportion of responding axillary meristems increases with the duration of water shortage. With a treatment of about 2 months, the formation of flower buds takes place only after the return of the plants to conditions of high watering while with a treatment of 3 months, flower initiation and development occur during the treatment itself. Thus, depending on its length, a period of water shortage may have an inductive or a direct action. Of greatest interest are the many similarities between these observations and those to be reported in Volume I, Chapter 5 concerning the effects of low temperatures.

Both water stress and chilling reduce growth and result in starch and protein hydrolysis with the consequence of an increased availability in soluble carbohydrates and amino acids, especially proline (see Volume II, Chapters 7 and 8).

There are reasons to believe that the case for *Geophila* is not exceptional and that "xeroinduction" is not restricted to species of equatorial origin. Support for this idea arises from the *in vitro* studies of Bouniols with root explants of *Cichorium intybus* (chicory), a biennial plant.[49] Meristems which are formed *de novo* on these explants flower readily provided they are submitted to LD and that sucrose is present in the culture medium (see Volume II, Chapter 1, Section II. A.3). However, they do so only if there is no excess water in their environment, either in the substrate (Figure 3) or in

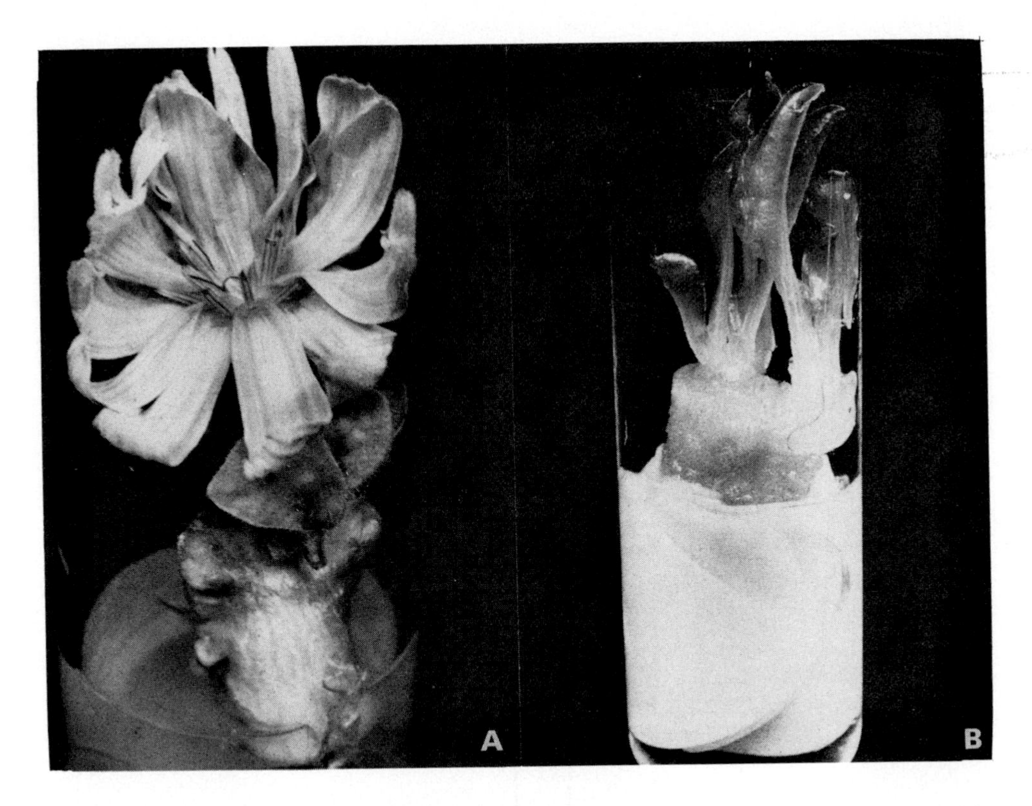

FIGURE 3. Morphogenetic potentialities of root explants of *Cichorium intybus* grown in continuous light. (A) Inflorescential bud formed on a solid (agar) medium; (B) vegetative buds formed on a liquid medium (the explant grows on a filter paper dipping in the medium). (Courtesy of A. Bouniols.)

the atmosphere (Figure 4). In the presence of excess water, these explants have a much greater water content and are able to regenerate vegetative buds, but their flowering is much depressed. Applications of water have also proved to be inhibitory to flowering in the SDP *Chenopodium polyspermum*.[50]

On the other hand, water stress during photoinduction in the SDP *Pharbitis* and *Xanthium* and the LDP *Lolium* prevents flower production, and there is evidence that this is due to a stress-imposed inhibition of translocation of floral stimuli from induced leaves.[51]

CONCLUSIONS

Mineral nutrition and water stress are generally believed to be of secondary importance in flower initiation. In many photoperiodic and cold-requiring plants, changes in the supply of N are unable to alter the inductive requirements. Thus the N supply in these species is only a modulating factor when other conditions are permissive of flowering. In a few other species, however, the process of flower initiation was found more directly tied to N availability. Trace elements, too, particularly copper and iron, are critically involved in photoperiodic induction in duckweeds and other plants. The mode of action of these two micronutrients is unknown.

Water availability is apparently an essential factor in the control of flower formation in some plants. Studies on this question have been so far very limited and should be extended.

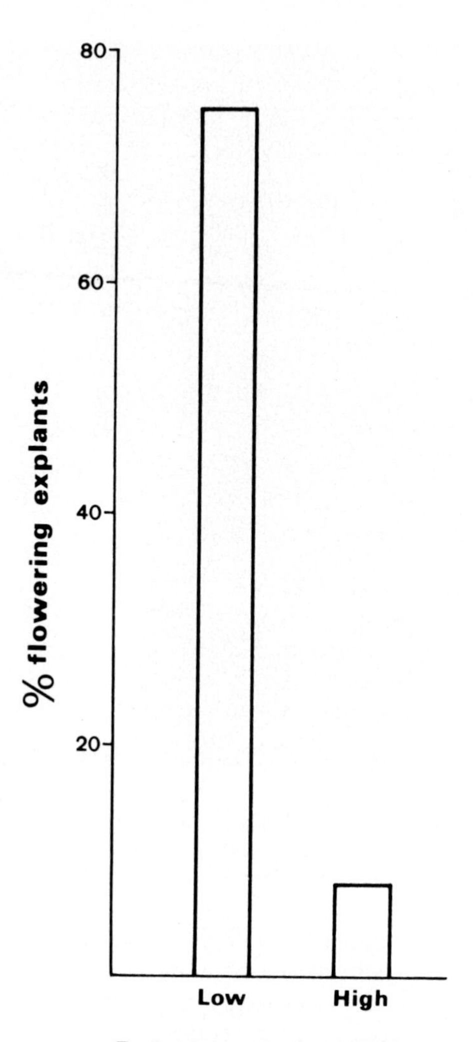

FIGURE 4. Flowering of meristems formed *de novo* in root explants of *Cichorium intybus* grown on a solid medium as a function of air humidity in the culture vessel. Cultures are kept in continuous light. (Adapted from Bouniols, A., *Plant Sci. Lett.*, 2, 363, 1974.)

Chapter 3

CONTROL BY DAYLENGTH

TABLE OF CONTENTS

I. INTRODUCTION

The history of the discovery of photoperiodism, well recounted elsewhere, [24,52] needs not be repeated. Suffice it to say here that the credit for this discovery is shared by Tournois who disclosed in 1914 that daylength is the controlling factor of the flowering of 2 SDP, hop and hemp, and by Garner and Allard who extended this observation between 1920 and 1940 to a wide range of field, garden, and ornamental plants and coined the terms "photoperiod" and "photoperiodism". In addition to the three basic response types described by these authors, viz., the SDP, LDP, and DNP, four others were recognized subsequently: the LSDP, SLDP, intermediate, and ambiphotoperiodic plants.

It has been repeatedly emphasized that the photoperiodic classification of plants has nothing to do with the particular daylengths at which the plants will flower, but whether flowering is promoted when the daylength is increased or decreased. Both the SDP *Perilla* and the LDP *Hyoscyamus,* for example, flower at daylengths between 15 and 13 hr (Figure 1).

In most photoperiodic response types, there are plants with an absolute response and others with a quantitative one. Thus plants referred to as *absolute* (or *qualitative*) SDP or LDP have an obligate requirement for SD or LD, respectively. These plants are characterized by an abrupt change in behavior over a narrow range of daylengths and consequently have a sharp "critical daylength" (Figure 1). On the other hand, *quantitative* (or *facultative*) SDP or LDP will produce flower buds under any daylength, but will do it earlier in SD or in LD, respectively. Such plants may or may not have a clear-cut critical daylength (Figure 1).

The concept of critical daylength is not without difficulties. Even in absolute photoperiodic species, the critical daylength can be profoundly modified by various environmental parameters, e.g., nutrition, temperature, light flux, by plant age, etc. Examples of conditional or operational photoperiodic behavior, i.e., behavior which is dependent on environmental or other conditions, will be given in subsequent sections.

A factor influencing the critical daylength is the number of favorable cycles given. In the LDP, *Lolium* and *Sinapis,* the critical daylength is several hours shorter in plants grown continuously under various daylengths than in those given only one LD cycle (Figure 2).[7,8] Similarly, the critical daylength is about 1 hr longer in *Xanthium* plants receiving five successive SD cycles than in plants receiving only one such cycle.[6] In general, plants submitted to repeated cycles appear to measure time more accurately, i.e., there is small variation in response between the individuals of a population, than when they are subjected to one or a few cycles.[6]

II. PHOTOPERIODIC INDUCTION; ITS QUANTITATIVE NATURE

It is a common observation that a brief favorable daylength regime is sufficient to bring about subsequent flower initiation even after the plant has been returned to unfavorable daylength conditions. Production of flower buds is thus an *after-effect* of the previous favorable photoperiodic treatment, which has accordingly been referred to as "photoperiodic induction" or "photoinduction". Plants which require only a single inductive cycle, for instance, never form flower buds during this cycle, but only a number of days after the return to noninductive conditions. In several SDP and LDP, the initiation of flowers starts at about the same time whether the plants have received a minimal or an optimal induction.[21] This is illustrated in Table 1 in the case of *Sinapis.*

Induction is clearly a progressive process with a well-marked quantitative nature. The case for the LDP, *Silene armeria,* is illustrative. This plant can be induced to

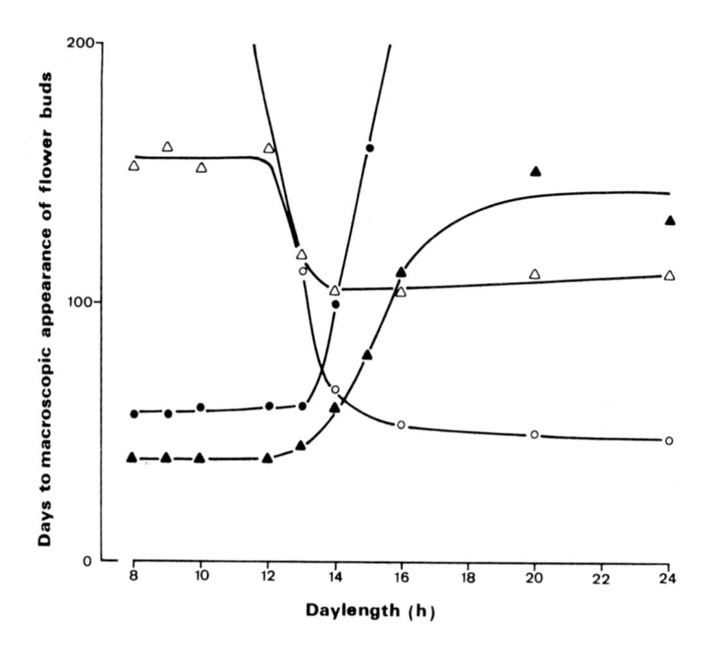

FIGURE 1. Days to macroscopic appearance of flower buds as a function of daylength in four different species grown at 20°C and in fluorescent light at the irradiance of 17 to 18 W·m^{-2}. The various daylengths are applied since sowing. ●, *Perilla nankinensis*, a qualitative SDP; ▲, *Cosmos variegata*, a quantitative SDP; ○, annual *Hyoscyamus niger*, a qualitative LDP; and △, *Coleus blumei*, a quantitative LDP. (Adapted from Bouillenne, R., *Bull. Cl. Sci. Acad. R. Belg.*, 5e série, 49, 337, 1963.)

flower in LD, but also in SD provided the temperature is raised from 20 to 32°C.[53] Four such warm SD cycles are below the threshold for induction, but when they are followed by two LD, which alone are marginally inductive, flowering occurs in a large majority of plants (Table 2). Thus, the effects of two apparently subthreshold treatments are additive, and the final result clearly depends on the level reached by each of the two treatments. In other words, various degrees of induction can be recognized even before a threshold permitting minimal flowering is reached. This phenomenon, called "fractional induction" will be explored more fully in a later section.

A similar quantitative response is commonly observed when the number of inductive cycles given exceeds the minimum required for causing flower formation. In many plants optimal induction, compared to a minimal one, results in production of more flowers or inflorescences (Table 1) and in an acceleration of the development of these reproductive structures. In *Lolium temulentum* the apex length, which is used to estimate the rate of inflorescence development, increases in proportion to the logarithm of the number of LD given (Volume I, Chapter 1, Figure 4). No discontinuity is apparent between the effect of the first LD, the only LD absolutely required for induction, and those subsequent to it even those given after spikelet and floret initiation has begun.[20]

It is thus quite possible to detect and measure degrees of photoinduction which are below or above the threshold necessary for the start of flower initiation. The change over to the initiation of the reproductive structures, either the flower or the inflorescence, occurs at some point during this progressive *quantitative* change in the determining factor(s).

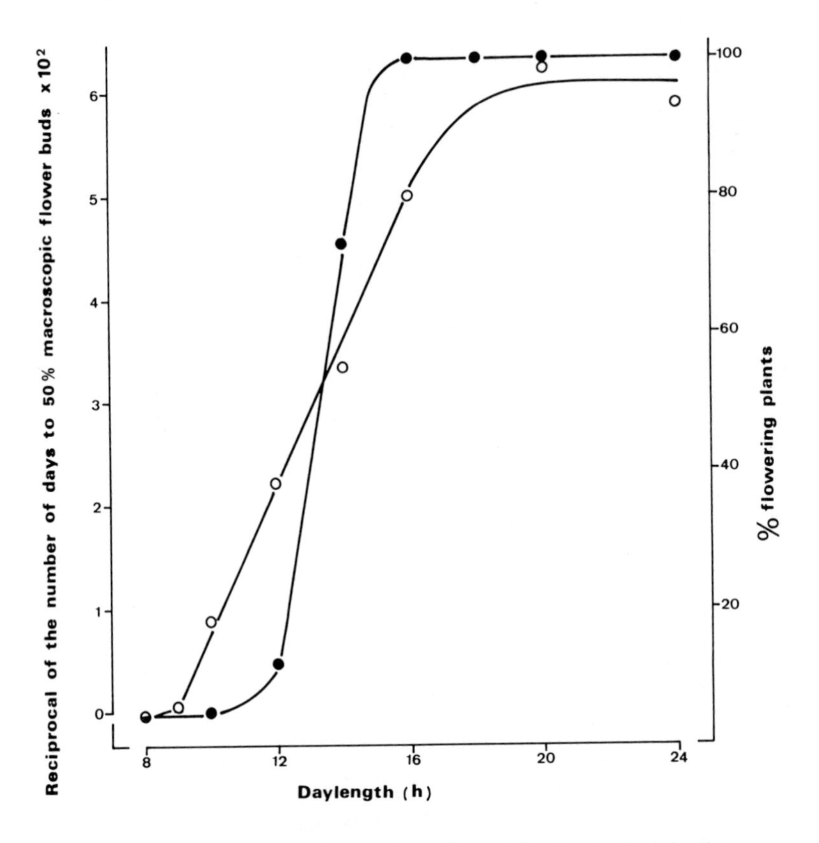

FIGURE 2. The response to daylength for flower initiation in *Sinapis alba* grown at 20°C under fluorescent light. O-O: various daylengths applied since sowing at the irradiance of 17 to 18 $W \cdot m^{-2}$; ●-●: a single day of increasing length given at the irradiance of 22 to 25 $W \cdot m^{-2}$ to 2-month-old vegetative plants grown in 8-hr days. The curves are sigmoidal suggesting a normal distribution of sensitivity of the population to daylength.

Table 1
NUMBER OF FLOWER PRIMORDIA PRODUCED BY THE APICAL MERISTEM OF THE LDP *SINAPIS ALBA* IN RESPONSE TO A MINIMAL OR A CONTINUOUS INDUCTION

Inductive treatment	Hr after start of the first LD				
	48	72	120	192	240
One 20-hr day	0	3	7	10	12
Ten 16-hr days	0	2	10	23	34

III. IMPORTANCE OF BOTH THE DARK AND THE LIGHT PERIODS

Work with the SDP *Xanthium* was clear-cut in indicating the crucial role of the dark period. Grown in 24-hr cycles, this plant flowers only when the dark period exceeds 8.5 hr. Hamner and Bonner asked whether the response was determined by the length of the dark period or by that of the light period.[54] With 24-hr cycles it is impossible to

Table 2
PERCENT OF FLOWERING PLANTS IN THE LDP *SILENE ARMERIA* (LINE S1.1) GIVEN SD AT 32°C AND LD AT 20°C

Number of SD cycles at 32°C	Number of LD cycles at 20°C			
	0	2	4	6
0	0	13	75	100
4	0	75	100	100
8	38	100	100	100
12	100	100	100	100

From Van de Vooren, J., *Z. Pflanzenphysiol.*, 61, 332, 1969. With permission.

vary independently the lengths of the two periods, but this is feasible when using light-dark cycles shorter and longer than 24 hr. They found that *Xanthium* does not flower in a 4-hr light per 8-hr darkness, but does in cycles of 16-hr light per 32-hr dark. The key is thus nightlength which must exceed a critical value. The same workers also discovered that a brief (1 min) light interruption in the middle of an otherwise inductive dark period nullifies the effect of this long night, i.e., the plants remain vegetative after such a treatment, while those in an uninterrupted night flower (Figure 3). An interruption of the day by a period of darkness has, on the contrary, no effect.

Rather brief light interruptions of long nights also inhibit flower initiation in several other SDP, e.g., Biloxi soybean, *Pharbitis, Perilla, Chenopodium rubrum, Lemna paucicostata* 6746, and *Kalanchoe,* the plants behaving as if they were in LD.[21,55] Night breaks promote flowering in some LDP, e.g., *Anethum, Hyoscyamus, Anagallis, Lolium, Brassica, Sinapis, Fuschia,* and barley (Figure 3).[21,55] Thus these plants flower when grown in a regime of SD provided the long nights are interrupted by light. However, as a rule, relatively brief night breaks are without any effect in LDP.[3,55-58] These plants usually require either night breaks of several hours and/or that the treatment is repeated during several successive nights.

Maximum effectiveness of night breaks is very dependent on their time of application during the dark period. It was initially observed that, under 24-hr cycles, the plants are usually most sensitive to light interruptions near the middle of the long nights. More refined observations have shown that the situation is not that simple and differs slightly from plant to plant. In *Xanthium* and *Pharbitis,* the time of maximum effectiveness is close to the 8th hr in a 16-hr dark period (Figure 3).[6,59] In *Coleus fredericii* and *Lolium,* strain Ba 3081, this time is after the middle of a 16-hr night.[55,60] These variations are probably related to differences in the mechanism of time measurement in different plants, a topic that will be considered in the next chapter.

These experiments with night breaks indicate clearly that the daily dark periods are of great importance in photoinduction in both SDP and LDP.

The daily light periods have an equally essential role. Using the SDP Biloxi soybean, Hamner investigated the effect of a combination of a constant near-optimal dark period of 16 hr (the critical dark period in this plant is about 10 hr) with a light period of variable length.[61] He found that flower initiation increases when the light period is lengthened from 4 to 11 hr and decreases when light is given for longer periods. Light

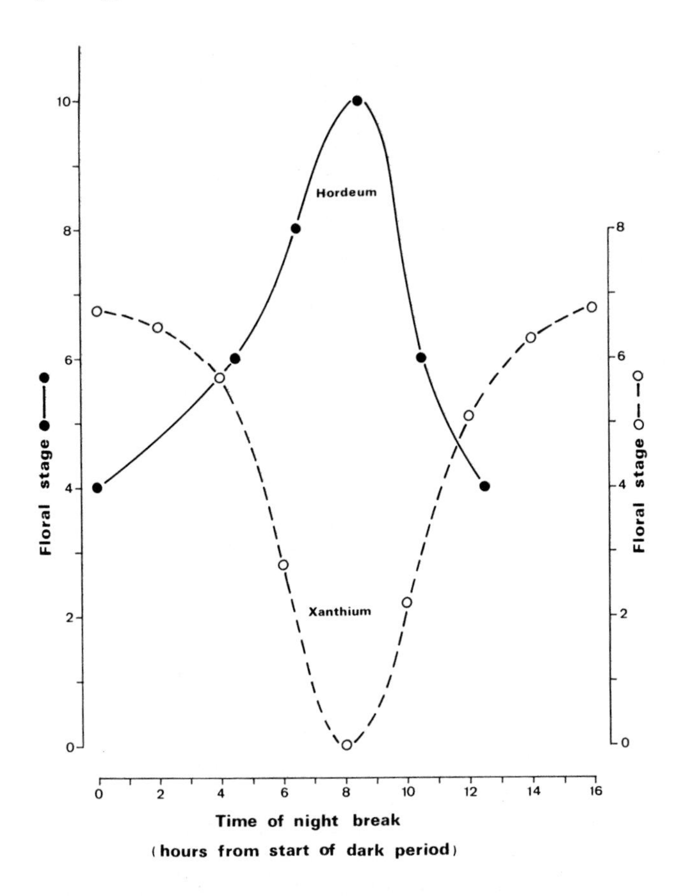

FIGURE 3. The effect of a night break given at various times during a long dark period in the SDP *Xanthium strumarium* and in the LDP *Hordeum vulgare* var. Wintex. *Xanthium* given a 16-hr dark period. *Hordeum* given a 12.5 *hr dark period.*[134,383]

periods of 20 hr or more totally inhibit flowering. These results show that flower formation in SDP is favored by an alternation of light and darkness.

LDP, on the other hand, flowers readily in continuous light and thus do not require darkness (Figures 1 and 2). They also flower when short light periods are combined with short, but not long, dark periods.[21] The inhibitory effect of a long dark period is totally overcome by light periods longer than a certain value. Thus these plants flower when exposed to cycles consisting of long light and long dark periods. Clearly, in both SDP and LDP, light and dark periods interact and the effect of a night (or day) of a certain length is dependent on the length of the associated day (or night). This kind of result is probably as it ought to be considering the importance of light-on and light-off signals in time measurement by plants (see Volume I, Chapter 4).

IV. FLOWER INITIATION IN COMPLETE DARKNESS

Many biologists entertain themselves by comparing their experimental material to a black box. Others take pleasure in growing their plant material *in* a black box, and the surprising outcome of this is an ever increasing list of plants, belonging to various photoperiodic response types, which are perfectly capable of initiating flowers in total darkness.[21] Examples are the SDP *Pharbitis*,[62] and the LDP *Sinapis, Arabidopsis,* and *Rudbeckia.*[63-65] Plants like sugarbeet and winter wheat, requiring a sequence of cold temperature and LD are also able to flower in the total absence of light after an appropriate vernalization.[66,67]

Except when the seed, or the root in the case of the beet, is rich in reserve material, the presence of sugar in the culture medium is a prerequisite for flower production in the dark. Remarkably, sugar beet and the *ld* mutant of *Arabidopsis*, which are normally obligate LDP, flower more rapidly in total darkness than in LD! In the LDP *Rudbeckia*, two suboptimal inductive treatments, the first with continuous darkness and the second with LD, are summated.[65]

In view of this, we are at first left with the impression that light as a rule is inhibitory to flower initiation in most types of plants and that this inhibition can only be overcome in natural conditions by appropriate light/dark cycles. However, among the plants that are capable of reproductive development in the total absence of light, some, e.g., the LDP, can also be brought into flowering by continuous illumination!

V. THE PERCEPTION OF DAYLENGTH

It is a very general rule that daylength is most effectively perceived by the leaves. This was first shown by Knott in the LDP *Spinacia*,[68] and rapidly confirmed and extended to a great number of species belonging to all photoperiodic response types.[21] Thus photoperiodic treatment of the shoot apices alone, where the process of flower initiation will be ultimately consummated, is ineffective; *only* leaves need to be exposed to favorable daylength conditions to achieve the induction of flowering. The direct experimental proof of this was given by Lona who induced *detached* leaves of the SDP *Perilla* and grafted them to vegetative plants grown in LD: these plants flower (see Reference 69). Leaves given LD are without such an effect. There is a close correlation between the minimal duration of SD treatment for flower initiation in intact plants of *Perilla*, on the one hand, and the capacity of leaves from these plants to function effectively as donors in grafting experiments, on the other.[70] Similar results are difficult to extend to other plants because isolated leaves usually do not survive long enough to be photoinduced. A partial success has however, been obtained in the LDP *Hyoscyamus* and in the SDP *Xanthium*.[21,71] Leaf discs from another SDP, *Streptocarpus nobilis*, grown *in vitro*, regenerate flower buds in SD and vegetative buds in LD, indicating that these small leaf explants can be successfully induced in an excised state.[72]

The sensitivity of leaves to photoinduction is dependent on several factors. First, sensitivity varies with physiological age. In most cases, maximal sensitivity is reached concurrently with full size. Very young leaves are usually far less effective than expanded ones. An exception to this is found in *Anagallis*, GO strain, in which leaves one-twentieth fully expanded are far more sensitive than half-expanded and older leaves.[56] In *Xanthium*, *Rottboellia*, and *Blitum*, peak sensitivity is attained at the time of most rapid expansion, when the leaf is about half-expanded.[73-75] Old leaves may have a declining efficiency, as in *Anagallis* and *Xanthium*,[56,73] or may retain their photoperiodic sensitivity for long periods, as in *Perilla* and *Lolium*.[71,76] The inductive capacity of leaves may also vary with their position on the stem, a topic which will be considered in Chapter 7.

That leaves are the primary receptors of daylength is clear from considerations about the minimal leaf area needed for photoinduction. As little as a few square centimeters from one sensitive leaf in *Xanthium* and *Lolium*,[73,76] a single leaflet of one young trifoliate leaf in soybean,[77] or a single cotyledon in *Brassica*,[78] are sufficient. Thus, while application of inductive photoperiods to the shoot tip, including the very young leaves, does not result in flower formation, but the same treatment of a minute leaf area is often sufficient to produce a significant response.

On the other hand, in several plants, the intensity of flower initation is related to leaf area, as in *Pharbitis*. Plants with only the first or second leaf do not react to a

Table 3
PERCENT FLOWERING[a] INDUCED BY A SINGLE LONG DARK PERIOD IN THE SDP *PHARBITIS NIL* AS A FUNCTION OF THE NUMBER OF LEAVES KEPT ON THE PLANTS

	Duration of the dark period	
Leaf present[b]	**18 hr**	**20 hr**
First leaf	0	17
Second leaf	0	30
First and second leaves	12	70

[a] Plants are disbudded before exposure to the dark period, except the axillary bud of the second leaf which is used as the "receptor" bud.
[b] Leaves present are removed at the end of the dark period.

Adapted from Imamura, S. I. and Takimoto, A., *Bot. Mag.*, 69, 289, 1956.

single long dark period of 18 hr and are weakly induced by a dark period 2 hr longer. If both the first and second leaves are kept on the plants, flower initiation is enhanced (Table 3). Necessity of a large leaf area in some species might not be related to photoinduction itself, but to the supply of a sufficient quantity of assimilates to the receptor meristems. In vitro, in the presence of carbohydrates in the culture medium, excised apices of several plants have been found to respond to photoperiodic treatments in essentially the same way as intact plants. Examples of this are the apices of the SDP *Cuscuta, Pharbitis,* and *Xanthium*[79-81] and of the LDP *Anagallis.*[82] Thus provided trophic and perhaps other materials are available, very small amounts of leaf tissue are sufficient for photoinduction.

Few SDP exhibit an unchanged photoperiodic behavior after complete defoliation demonstrating that daylength can be perceived in these cases by the stem or the shoot apex.[21,83] In vitro cultures of internode sections of *Plumbago indica* and of root explants of *Cichorium intybus* indicate that these plant pieces can respond to daylength in the total absence of leaves.[84,85] In view of all of this, it is fair to conclude that photoperiodic sensitivity is a property of all plant parts, but that this property is most developed in the leaves.

VI. FRACTIONAL PHOTOINDUCTION

In plants requiring more than one inductive cycle, these cycles need not always be given consecutively. Thus, interpolation of a number of noninductive cycles between two groups of inductive cycles which are each insufficient to bring about the floral transition does not prevent summation of the two subthreshold inductive treatments. This phenomenon, called "fractional induction", has been found in various LDP and SDP. It is easier to observe in some plants than in others and is markedly influenced by experimental parameters, such as when the intercalated cycles are given and their number, light, and temperature conditions during these cycles, etc. A few examples illustrate the kind of results obtained in these experiments.

Summation of nonconsecutive cycles does occur in plants of *Silene armeria* alternately exposed to one LD and one SD for 6 to 8 weeks.[53] In annual beet, a plant

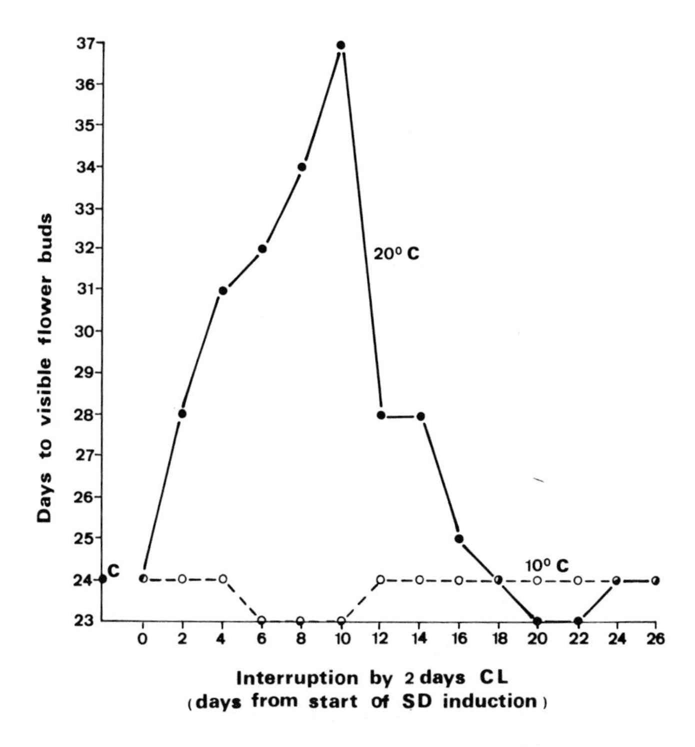

FIGURE 4. Effect on flower formation in *Salvia occidentalis* of a 2-day interruption by continuous light (CL) given at different times during the SD inductive treatment. CL interruptions are given at 20 or 10°C. C on ordinate indicates the uninterrupted SD control at 20°C. (From Bhargava, S. C., *Meded. Landbouwhogesch. Wageningen*, 64 (12), 1, 1964. With permission.)

requiring a minimum of 15 to 20 inductive LD cycles, flower initiation does not occur with long-term repetition of regimes consisting either of one LD followed by one SD or of three consecutive LD followed by three consecutive SD,[86] but two groups of ten consecutive LD separated by sixteen SD are summated.[61] No explanation has been advanced to account for persistence of the LD effect in the latter regime or its lability in the former.

Schwabe has described a rather similar situation in various SDP. In *Perilla*, for example, nine consecutive SD are required for 100% flowering. When twelve SD are given, with one LD in between each SD, the plant remains vegetative. Intercalating two LD in the middle of twelve SD does not prevent flower formation, but intercalation of three LD does. The effect of partial induction by six SD cycles is not carried over a period longer than 2 days in this case.[87]

In *Perilla*, as in several other SDP, the effect of interpolated LD depends markedly upon their position in the SD sequence. Thus, 2 days of continuous light most effectively inhibit flower formation in *Salvia occidentalis* when given after the tenth SD, just before the minimal number of inductive cycles is reached (Figure 4). Before and after the 10th day, the 2-day continuous light "interruption" inhibits much less. In addition, the inhibition caused by the interruption is temperature-dependent and does not occur at low temperature (Figure 4).

The situation in the SDP *Rottboellia* is still more complex.[74] One LD interpolated in a sequence of six SD promotes inflorescence initiation when given between the first two SD and inhibits it strongly when given later in the sequence.

In plants with axillary flowering, e.g., Biloxi soybean and *Anagallis*, fractional induction seems totally impossible. Two induction periods act independently in these species and subthreshold inductive treatments are apparently never summated.[56]

Obviously, we are facing an extremely complex situation in the field of fractional induction which seems to preclude any unitary interpretation. It is often assumed that a subthreshold inductive treatment causes no change at the apex, and summation necessarily occurs in the leaves. While this might be true, it would be quite unwise to accept that idea without appropriate investigations. Indeed, recent results, to be described in Volume II, Chapters 2 to 5, have indicated that part of the changes typical of evocation may occur in several species in response to subminimal inductive treatments (partial evocation).

If subthreshold inductive treatments to different sets of leaves can be summated, one would have indirect evidence that induction is summated at the shoot apex and not in the leaves. In *Pharbitis,*[1] *Xanthium,*[88] and *Perilla,*[70] subthreshold inductive treatments to different leaves are not summated, but a contrasting situation was described in *Rottboellia,* another SDP, under some circumstances.[74] In the LSDP, *Cestrum nocturnum,* Sachs showed that only leaves which had previously received the LD part of induction respond to SD.[89] In the other LSDP, *Bryophyllum daigremontianum,* the situation is far less clear. Chailakhyan and Yanina have grafted a stock continuously kept in LD and a scion continuously kept in SD and have obtained flowering in receptor shoots positioned between the LD and the SD partners.[90] This result seems to indicate that the LD and the SD can be perceived in this species by different leaves. However, Van de Pol was unable to reproduce this experiment,[91] and doubts have been raised concerning the possibility of spatial fractionation in *Bryophyllum.*[92] More work is necessary to settle this question, but if the evidence is considered in its entirety, such a fractionation does not seem possible in most investigated plants.

VII. PERMANENCE OF THE PHOTOINDUCED STATE

Interest in this problem arose as a consequence of the observation that in many photoperiodic plants flower initiation is an after-effect of exposure to favorable light/dark cycles, and thus that the effect of these cycles has been maintained for a certain duration of time. The so-called fractional induction experiments, when successful, also provide some evidence that the effects of inductive treatments that are not in themselves sufficient to bring about the transition to flowering may be conserved during a more or less extended period in noninductive conditions.

Close examination of this problem has, however, revealed that in the great majority of plants photoinduction, *when suboptimal,* is short-lasting and that the plants revert to vegetative growth as soon as the effect of this induction is consummated. Most often vegetative growth resumes in apices that are not converted to the reproductive condition as a result of a previous inductive treatment. In maximally induced annual or monocarpic species, resumption of vegetative growth is as a rule totally impossible because all responsive apices have been irreversibly oriented towards the production of flowers.

The induced state is apparently very persistent in a few remarkable species which, because of this property, have been preferentially selected in investigations on the induced state.

Fragaria vesca var. *semperflorens,* worked out extensively by Sironval,[93] has been held for 3 years from sowing in SD without any sign of flowering at all. However, plants induced to flower by exposure to LD do not return to normal vegetative growth after a transfer to 8-hr days for several months. New flowers are continuously produced, although in less number per plant, until the plant death about 10 months after start of the SD treatment.

After minimal induction, the progress of the SDP *Xanthium* towards flowering and fruiting is slow, but it continues for months. The plants ultimately revert to a vegetative condition, but this reversion takes place, in some shoots only, 6 months after induction.[94] More rapid reversion can be obtained by repeated and rather drastic disbudding of the plants forcing new shoots to grow out continuously.[94]

This stability of the induced state in *Xanthium* is perhaps related to the curious property of "indirect induction" exhibited by this plant, a phenomenon first disclosed by Lona[94a] and later investigated by Zeevaart.[71]

If a flowering *Xanthium* plant is grafted to a vegetative receptor plant, the latter flowers. If this indirectly induced *Xanthium* plant is in turn grafted to a new vegetative receptor, the latter also flowers and may in turn serve as a donor to another vegetative partner. In this way, the induced state can be transmitted through several successive grafts without any sign of a decrease of the flowering response. Thus the induced state, whatever it is, is transferable to plants or plant parts that have never experienced inductive conditions, hence the term "indirect,"[69] or "nonlocalized" induction.[91]

The impression gained in these experiments with *Xanthium* is that the floral condition is contagious, resembling a virus disease.[95] Accordingly, it has been proposed that the floral stimulus either reproduces autocatalytically,[53] or stimulates its own synthesis by a positive feedback mechanism in growing tissues, either bud or leaf.[92]

Three other species only, the LSDP *Bryophyllum daigremontianum,*[96] the LDP *Silene armeria,*[97] and the SDP green-leaved *Perilla,*[98] are known to share the property of indirect induction with *Xanthium*. Contrary to *Xanthium,* however, suboptimally induced plants of *Bryophyllum* and *Perilla* have been observed to revert to vegetative growth simply by returning them in unfavorable conditions.[91,99]

Although reversion to vegetative growth is also common in suboptimally induced individuals of the red-leaved *Perilla,* Zeevaart has found that an induced leaf of this species, grafted successively to several, up to seven, receptor plants kept in LD causes all of them to flower.[71] Even in the last grafting, just before senescence of the donor leaf, there is no decrease in flowering response of the receptor. On the other hand, flowering shoots or leaves of receptor plants do not function as donors for vegetative partners. Thus, the red-leaved *Perilla* does not exhibit the phenomenon of indirect induction. The induced state in this *Perilla* is apparently as persistent as in *Xanthium,* but is localized to the leaves that have been directly exposed to the inductive light/dark cycles.

How then do we explain the natural reversion of suboptimally induced plants of the red *Perilla* even in the presence of induced leaves? Lona and Zeevaart propose that the young leaves developing at the top of the stem after the return to LD are not induced, and they gradually take over control of the apical meristem.[69,99] However, by repeated disbudding (as in *Xanthium*), Lam and Leopold demonstrated that *Perilla,* brought into flowering by exposure to only 10 to 20 SD cycles, reverts to the vegetative phase.[100] Plants receiving 27 SD cycles revert rarely, and those receiving 50 c never revert. Reversion in this study seems related to the decline in production of the floral stimulus which is an inverse function of the number of inductive cycles. This interpretation is disputed by Zeevaart on the basis of his finding that suboptimally induced leaves retain their induced state for a period equal to that of maximally induced leaves.[99] The situation, postulated by Lam and Leopold, i.e., the loss of a factor essential for differentiation or morphogenesis, may not be uncommon, however, when plant cells are forced to divide rapidly. Continuous removal of flower buds on floriferous stems of *Geum* results, over a period of 1 year, in an extraordinary multiplication of minute shoot meristems, up to 10,000 per floriferous stem, and a progressive loss of the vernalized state.[31] Similarly, tissue explants, excised from reproductive axes, may regenerate only vegetative buds. This has been observed in Maryland Mammoth tobacco, *Nicotiana sylvestris, Streptocarpus,* horseradish, carrot, and beet.[72,101,102] In day-neutral varieties of tobacco, *Cichorium, Begonia, Sinapis,* and *Lunaria,* pieces of flowering stalks can produce flower primordia in vitro,[101,103,104] but even in these cases Konstantinova and co-workers showed that the capacity to produce flowers decreases upon repeated passages in vitro. In inflorescence segments of the DNP Trapezond tobacco, cultures producing flower buds decline from 74 to 40% from the first to the third passage, suggesting progressive dilution of one or several essential compounds.[105] The tumorous state itself may be reversed by forcing tumor shoots of tobacco into very rapid but organized growth as a result of a series of tip graftings to healthy stocks.[106]

In all these cases, the simplest explanation is that one or several necessary "differentiation" factors present in limited amounts at the start of the experiment have been consumed or diluted during intense cell proliferation. The disappearance of the corresponding state of differentiation is the ultimate consequence.

VIII. INTERACTIONS OF DAYLENGTH AND OTHER ENVIRONMENTAL FACTORS

A. Temperature

The modifying influence of temperature on photoperiodism has been known even since the days of Garner and Allard. The fact that photoperiodic species do not flower in nature at an identical date year after year is usually attributed to differences in temperature. A classical example of this influence is that of annual *Hyoscyamus* worked out by Lang and Melchers.[21] In this LDP, the critical daylength shifts from 8.5 to 11.5 hr when night temperature increases from 15.5 to 28.5°C. Thus this plant would flower in shorter daylengths at lower temperatures.

There are other dramatic cases in which the photoperiodic response is completely changed by temperature. Many so-called "absolute" SDP, such as Maryland Mammoth tobacco,[107] cultivated strawberry,[108] *Perilla,*[98,99] *Pharbitis,*[59] and *Begonia* (Figure 5), require SD only when the temperature is in the range of 20 to 25°C, and eventually produce flowers in LD or even in continuous light at 15°C or below. It is not always clear which temperature, night or day, is important in these experiments because constant conditions are often used. However, there is evidence that low temperature affects dark processes in tobacco and photoprocesses in *Perilla.* The SDP *Xanthium* flowers in LD provided the temperature is around 4°C during the first half of the long light periods.[109]

The LD requirement can be completely replaced by a low temperature treatment in the LDP *Silene,*[53] *Blitum,*[75] *Anagallis,*[56] *Calamintha,*[110] *Melandrium,*[91] and *Sinapis.*[13] Interestingly enough, the substitute cold treatment is apparently perceived by the leaves in the SDP *Perilla* and *Pharbitis,*[98,111] and the LDP *Blitum.*[75]

In *Scabiosa succisa, Campanula medium,* and *Coreopsis grandiflora,* three SLDP, the requirement for SD can be totally overcome by a period of low temperatures.[112-114] A very long chilling treatment can even suppress the requirement for LD in *Campanula.*

Temperatures above 30°C may also completely suppress the requirement for LD in some LDP as first reported by Murneek in *Rudbeckia bicolor.*[115] *Silene armeria* and *Calamintha officinalis* also behave as LDP at moderate temperatures and DNP at high temperature.[53,110] According to Wellensiek, the high temperature treatment is perceived by the roots in *Silene.*[116]

B. Light Intensity

Examples of interactions between the duration of the light period and photon flux density are numerous among photoperiodic plants.

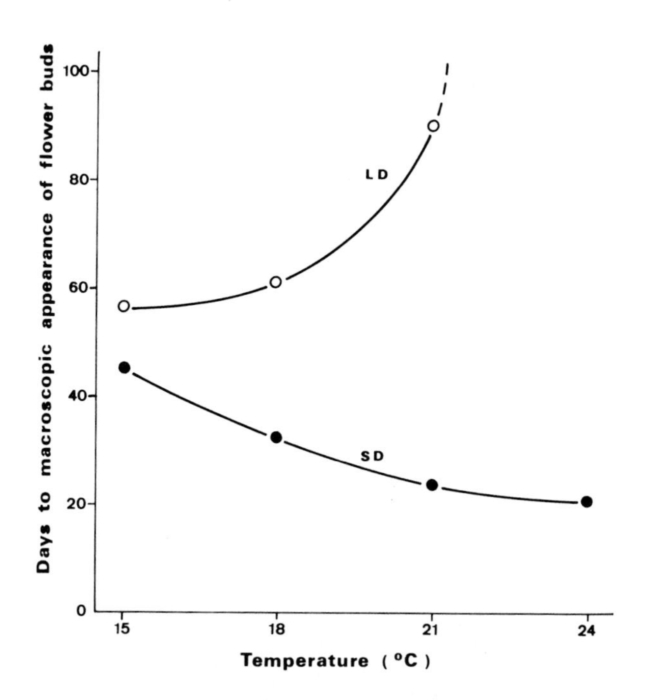

FIGURE 5. Effect of temperature on the photoperiodic response of *Begonia × cheimantha* cv. Mörk Marina. Note that this cultivar is an obligate SDP above 21°C and is almost indifferent to daylength at 15°C. (From Heide, O. M., *Z. Pflanzenphysiol.*, 61, 279, 1969. With permission.)

In the SDP Biloxi soybean exposed to 5- or 10-hr SD at light intensities varying from 500 to 8,000 lx, Hamner found that flowering is a direct function of light intensity above 1,000 lx. Below this value no flower initiation occurs.[61] *Xanthium* plants exposed to 12 c of 3 hr of darkness and 3 min of light for 36 hr do not flower when given an inductive dark period, i.e., they cannot be induced. A long period of low intensity light has the same effect as the 36 hr light/dark treatment. Restoration of full responsiveness to an inductive long dark period requires prior exposure of plants to light of a certain duration and intensity (Figure 6).

Changes in photon flux density may have a very striking effect in other SDP. Very low light intensities allow flower formation to proceed in LD or even in continuous light in *Perilla*,[117] *Salvia occidentalis*,[118] and *Lemna paucicostata* 6746.[119] High photon flux density may also result in flower production in continuous light in the "obligate" SDP *Pharbitis* (Table 4) and *Lemna paucicostata* 6746 in conjunction with other factors.[120]

Usually LDP are induced to flower by extending the SD given at high intensity light by a period of supplementary light at low irradiance, i.e., light below the photosynthetic compensation point. However several LDP, e.g., *Anagallis* and *Brassica*, do not respond or respond very poorly to LD of this sort. In these plants supplementary illumination at high photon flux density is essential for LD induction.[78,82] As shown by Bodson and associates, the flowering response of *Sinapis* to a single 16-hr LD is also very much influenced by the irradiance.[121] During the second half of the LD, very high and very low irradiances (above 96 and below 15 $W \cdot m^{-2}$, respectively) inhibit flower initiation while intermediate irradiances (25 $W \cdot m^{-2}$) promote it. Very high irradiances have, however, a promotive effect when applied during the SD preceding or following the LD, even during the first part of the LD (Figure 7). The effect of high light is thus strongly time-dependent in this species.

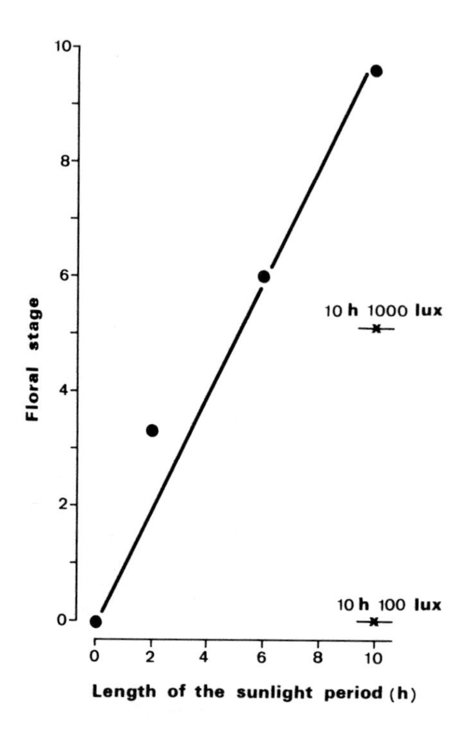

FIGURE 6. Effects on flower initiation of the SDP *Xanthium strumarium* of the length of a sunlight period or the intensity of a 10-hr light period inserted between 12 c, each consisting of 3 hr of darkness, 3 min of light, and a 12-hr inductive dark period. The original data were converted roughly to the floral stage system of Figure 2A in Volume I, Chapter 1. (Adapted from data of Hamner, K. C., *Bot. Gaz. (Chicago)*, 101, 658, 1940.)

Table 4
EFFECT OF LIGHT INTENSITY ON FLOWER INITIATION AND STEM GROWTH IN THE SDP *PHARBITIS NIL* UNDER CONTINUOUS ILLUMINATION

Light intensity (1x)	% of plants with flower buds	Mean number of flower buds per plant	Stem length (cm)
3,000	0	0	64.6
6,000	0	0	57.1
9,000	0	0	60.0
16,000	100	3.6	6.0
26,000	100	3.1	6.5

From Shinozaki, M., *Plant Cell Physiol.*, 13, 391, 1972. With permission.

High photon flux densities during a regular SD regime may even completely obviate the need for LD in *Sinapis*,[8,121] *Brassica*,[122] and other LDP. This effect of high light

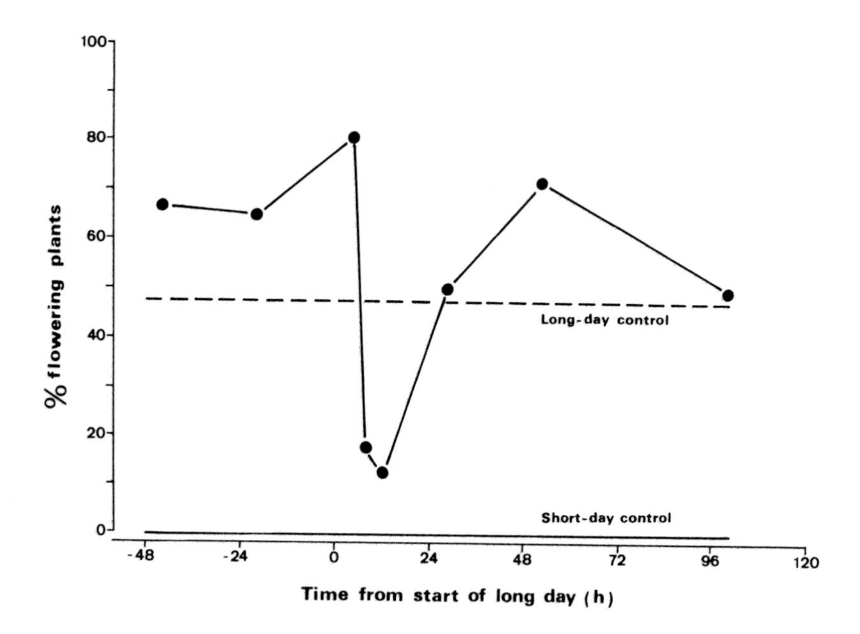

FIGURE 7. Flower initiation in *Sinapis alba* as a function of an 8-hr period of high irradiance (96 $W \cdot m^{-2}$) imposed at various times before, during, or after exposure to one 16-hr LD at an irradiance of 25 $W \cdot m^{-2}$. Plot points are the midpoint of the time of exposure. (From Bodson, M., King, R. W., Evans, L. T., and Bernier, G., *Aust. J. Plant Physiol.*, 4, 467, 1977. With permission.)

receives further attention in Volume II Chapter 8. High irradiances may, on the contrary, inhibit flowering in *Lemna gibba* grown in LD.[120]

C. Atmosphere Composition

Removal or addition of atmospheric CO_2 can alter the photoperiodic response of many plants. This problem was fully discussed in a recent symposium.[123] Suffice it to say here that CO_2 removal during the light period reduces flower initiation in several SDP and LDP.[21,63,124,125] In LDP, CO_2 is apparently not required during the hours of supplementary illumination, but only during the first part of the LD (Figure 8).[21,63] Carbon dioxide enrichment, on the other hand, may suppress flower formation in inductive conditions in the SDP *Lemna paucicostata* 6746,[126] *Xanthium*, and *Pharbitis*,[127] and in the LDP *Lemna gibba*.[120] In the LDP *Silene*, 1 to 1.5% CO_2 promotes flowering in plants held in SD.[127]

CO_2 withdrawal prevents the effects of light interruptions (night breaks) of extended dark periods in the LDP *Brassica* and the SDP *Xanthium*, but not in the SDP *Perilla*.[128,129]

Several LDP have been caused to flower in SD by holding them in an atmosphere of nitrogen during the dark periods. This finding was first reported with *Hyoscyamus* and later extended to *Rudbeckia* and *Lolium*.[21,130] Conversely, in several SDP, e.g., *Perilla* and soybean, the response to SD is nullified by a nitrogen atmosphere during the long nights, particularly if given in the second half of these nights.[131]

IX. RECONSIDERATION OF THE CLASSIFICATION OF RESPONSE TYPES

So far, only simple interactions, i.e., those involving daylength and one other factor, have been discussed. Extremely complex situations develop when more factors are con-

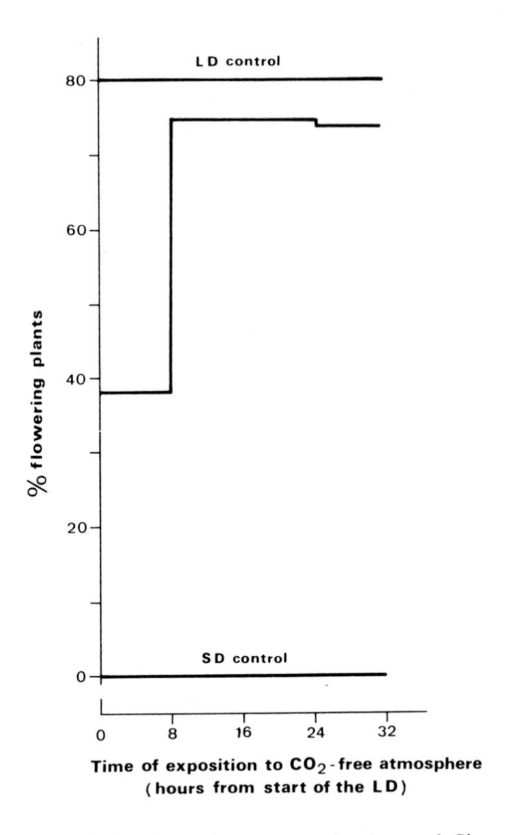

FIGURE 8. Flowering response in plants of *Sinapis alba* exposed to a single 16-hr LD as a function of time of exposure to a CO_2-free atmosphere for 8 hr. (From Kinet, J. M., Bernier, G., Bodson, M., and Jacqmard, A., *Plant Physiol.*, 51, 598, 1973. With permission.)

sidered simultaneously. In one experiment, Went exposed Marshall strawberry plants to 8 hr sunlight at 23°C plus various combinations of temperature and irradiance during the remaining 16 hr.[108] This plant is a typical SDP at moderate temperatures and a DNP at 10°C or below (see Volume I, Chapter 3, Section VIII. A). In continuous light, flower initiation is totally suppressed at 17°C by light in the 4000- to 12,000-lx range, but not at 10°C at any intensity within this range. At 14°C, flower initiation occurs at 4000 lx, but not at 8000 lx and above. In other words, the lower the temperature, the higher the light intensity has to be to prevent flower formation in LD. Thus, strawberry is considered a DNP at low temperatures only because the irradiance which will suppress flowering is probably higher than was easily attained in growth cabinets.

Another example is that of *Lemna paucicostata* 6746 which behaves as an obligate SDP when grown in Hoagland's medium totally devoid of cupric ion and as a DNP in the same medium supplemented with this ion (Volume I, Chapter 2). Even with Cu^{2+}, however, indifference to daylength is only found at 28°C or below. At higher temperatures, say 29.5°C, this plant only flowers in SD.[5]

These few examples show convincingly that all environmental factors may interact such that each factor may change the threshold values of the others.

We are far now from the appealing simplicity of the principles discovered by Garner and Allard. The critical daylength, the cornerstone of classification of plants according to their photoperiodic response, is far from constant in all conditions. In its entirety the evidence reviewed demonstrates amply that the photoperiodic response of plants is not only modulated, but also profoundly altered by changes in various environmental parameters. Classification of a plant in one or another photoperiodic group is only valid under a given set of conditions, and it may be completely modified in other conditions.

Moreover the preceding studies reveal the fantastic diversity in response of plants to their environment, a point repeatedly made by Chouard,[132] which is evidently related to their adaptation during evolution. In the light of this, one is tempted to introduce a number of refinements in the classification in order to cope with the variety of plant responses.[6,55,112] However, even very complex classifications, including dozens of response types, cannot cover the variety of responses known, and they will become less and less satisfactory as experimental work goes on.

Perhaps the main interest in classifying responses of plants to the environment is for the ecologist seeking explanations for plant geographic distribution. But for the physiologist, classification without further qualification is not very helpful because it gives no insight into the *endogenous control systems* and the intimate nature of the process of flower initiation.

CONCLUSIONS

Daylength is one of the major environmental factors controlling flower initiation in many plants. The photoperiodic response of plants is not a fixed property, however, and is amenable to considerable variations. Examples of profoundly altered responses obtained by changing temperature, photon flux, or atmospheric composition have been presented. Thus, classification of a plant in one or another response group is only valid under a given set of conditions.

Photoperiodic sensitivity is a property of all plant parts, but as a rule, daylength is most effectively perceived by the leaves. In many photoperiodic plants, initiation of flowers is an after-effect of exposure to favorable light/dark cycles. This induction, when suboptimal, is transient in most plants, and reversion to vegetative growth is then observed as soon as its effect is consummated. In a few species the induced state is more persistent. Whatever the mechanism(s) of this persistence, and there are apparently important differences among species, these plants also revert more or less rapidly to a vegetative condition after suboptimal induction. More rapid reversion is usually obtained by repeated disbudding of the plants forcing new shoots to grow out continuously. Similarly, in pieces from reproductive axes of many plants cultured in vitro, the capacity to regenerate flower buds may be lost after the first passage or may decrease upon repeated passages. In all cases, the factor(s) responsible for the induced state are apparently consumed or diluted more or less rapidly during intense cell proliferation.

Chapter 4

LIGHT PERCEPTION AND TIME MEASUREMENT IN PHOTOPERIODISM

TABLE OF CONTENTS

I. LIGHT PERCEPTION

The perception of light in photoperiodism is essentially attributed to the pigment phytochrome and, insofar as flowering is concerned, this is possibly the only photoreceptor.

A. History

Early work leading to the discovery of phytochrome is closely associated with studies on flowering and we will focus only on these studies, although present-day research is primarily on many other plant responses, more amenable to investigation of phytochrome action.

Soon after Hamner and Bonner found that a brief exposure to white light in the middle of an inductive dark period prevents flower formation in *Xanthium* (see Volume I, Chapter 3, Section III),[54] Parker et al. established the action spectra for the effect of such light breaks in two SDP, *Xanthium* and Biloxi soybean (Figure 1).[133] They observed that the most effective region of the visible spectrum in suppressing flowering is a rather broad band in the red, extending from 600 to 680 nm, with no effect beyond 720 nm. A second, but less active, region is in the blue near 400 nm. Minimum effectiveness occurs near 480 nm (green). Though the curves are qualitatively the same for both species, there are quantitative differences in the energy required at certain wavelengths. For instance, at 480 nm, *Xanthium* requires more than seven times as much energy as soybean. In the region of maximum effectiveness the energy required per unit of response is nearly the same in both species.

Some years later, action spectra for light interruptions of noninductive long dark periods on the flowering response of the LDP Wintex barley (Figure 1) and annual *Hyoscyamus* were determined.[134,135] In the red region, the response is similar both qualitatively and quantitatively to that for SDP. In the blue region, no clear response is detected in *Hyoscyamus,* and in barley more energy is required to initiate spike development than to suppress flower initiation in the SDP. These similarities in action spectra suggest that the same photoreceptor is involved in the control of flowering in both SDP and LDP.

In 1952, Borthwick et al. demonstrated that the red light inhibition of flowering in *Xanthium* can be reversed by a following far-red irradiation treatment.[136] The action spectrum for repromotion of flowering shows that maximum effectiveness is near 730 nm. The photoreversibility of flower production was later extended to two SDP, Biloxi soybean and *Amaranthus caudatus* and two LDP, Wintex barley and *Hyoscyamus.*[137] Furthermore, by subjecting *Xanthium* and soybean plants to several consecutive irradiations with red and far-red in sequence, Downs observed that the reaction is repeatedly reversible.[137] With both species, the flowering response is determined almost entirely by the wavelength of the final exposure (Table 1). However, the degree of repromotion becomes less with each red/far-red cycle, until finally after several such cycles, repromotion is no longer evident.

From these experiments, Borthwick and associates deduced that the photoreceptor, that they called "phytochrome", occurs as two interconvertible forms, one with an absorption maximum in the red region of the visible spectrum (Pr or P_{660}), the other predominantly absorbing far-red radiation (Pfr or P_{730}). [136] Red radiation (R) would convert the pigment to the Pfr form, and conversely, far-red light (FR) would change the pigment to Pr. White light acts like R. Pfr, produced in response to R, is assumed to be the physiologically active form since relatively small amounts of Pfr cause a response. A simple induction/reversion model of phytochrome action can thus be formulated as follows:

$$P_r \underset{FR}{\overset{R}{\rightleftharpoons}} P_{fr} \longrightarrow \text{photoresponse}$$

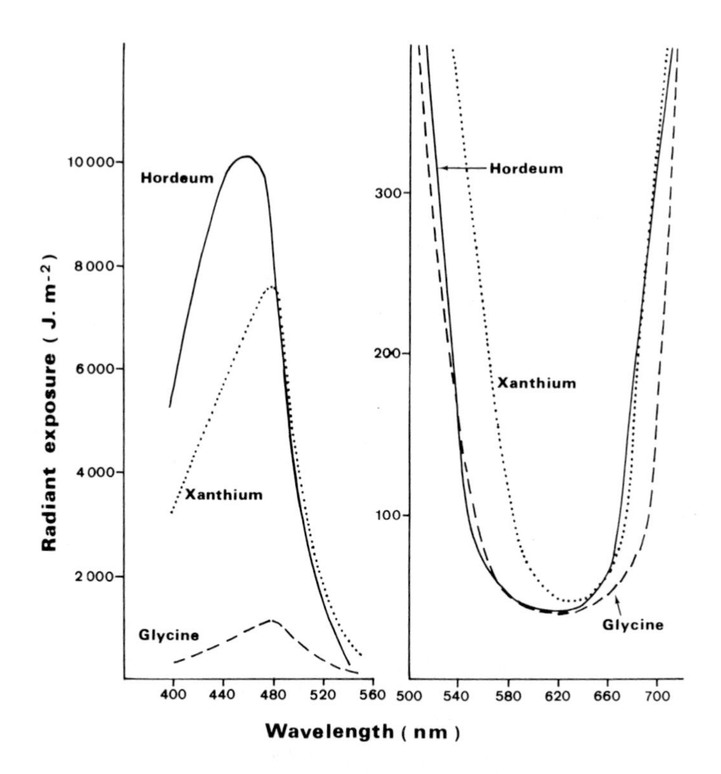

FIGURE 1. Action spectra for night-break effects on floral initiation of the SDP *Xanthium strumarium* and *Glycine max* and the LDP *Hordeum vulgare*. For SDP: energy required to suppress floral initiation. For the LDP: energy required to induce floral initiation. (From Borthwick, H. A., Hendricks, S. B., and Parker, M. W., *Bot. Gaz. (Chicago)*, 110, 103, 1948. With permission.)

Using R/FR reversibility as a criterion, numerous plant responses to light have been shown to be controlled by phytochrome. Generally, the radiant exposure with R needed to saturate phytochrome responses is reduced: it is about 30 $J \cdot m^{-2}$.

B. Molecular Properties of Phytochrome

The pigment was first extracted from etiolated oat coleoptiles in 1959.[138] It is a chromoprotein which natively occurs as a multimer of 120,000 dalton subunits.[139] The chromophore is a linear tetrapyrrole similar to C-phycocyanin.[140]

Phytochrome is widely distributed in the plant kingdom. High levels are present in all etiolated angiosperm seedlings and it has been extracted (or detected by in vivo spectrophotometric assay) from all parts of higher plants. However the phytochrome content of green leaves, the organs which perceive daylength most effectively (see Volume I, Chapter 3, Section V is very low and this pigment has not yet been detected in *Xanthium*, soybean, and *Perilla* (among the photoperiodically most sensitive species). The highest levels of phytochrome are usually found in meristematic tissues.

Since phytochrome is detected with difficulty in green tissues, most studies use etiolated seedlings. Highly purified preparations are obtained from dark-grown monocotyledons, but several investigations in vitro utilized degraded phytochrome and thus should be repeated with the undegraded form.[139]

The absorption spectra of the two forms of phytochrome isolated reveal considerable overlap (Figure 2),[141] making total photochemical conversion impossible, regardless of wavelength used. The proportion of Pfr to total phytochrome (Pfr/P), at photo-

Table 1
EFFECT OF SUCCESSIVE RED (R)/FAR-RED (FR) CYCLES GIVEN IN THE MIDDLE OF AN INDUCTIVE DARK PERIOD ON FLOWER INITIATION IN THE SDP *XANTHIUM STRUMARIUM* AND BILOXI SOYBEAN

Treatment[a]	Floral stage in *Xanthium*	Number of flowering nodes in soybean
Dark control	6.0	4.0
R	0.0	0.0
R,FR	5.6	1.6
R,FR,R	0.0	0.0
R,FR,R,FR	4.2	1.0
R,FR,R,FR,R	0.0	—
R,FR,R,FR,R,FR	2.4	0.6
R,FR,R,FR,R,FR,R	0.0	0.0
R,FR,R,FR,R,FR,R,FR	0.6	0.0

[a] For *Xanthium*, both R and FR of 2-min duration; three 12-hr dark periods. For soybean, R for 2 min, FR for 8 min; nine 14-hr dark periods.

From Downs, R. J., *Plant Physiol.*, 31, 279, 1956. With permission.

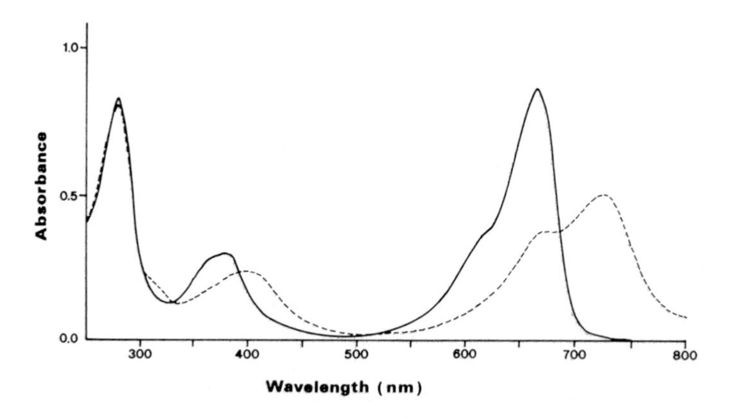

FIGURE 2. Absorption spectra of Pr (———) and Pfr (- - -) forms of isolated oat phytochrome after saturating irradiation with FR or R light. (Reproduced with permission from Mumford, F. E. and Jenner, E. L., *Biochemistry*, 5, 3657, 1966, Copyright 1966 by The American Chemical Society.)

chemical equilibrium, is about 75% at 660 nm (photostationary state = 0.75) and usually less than 3% at 730 nm (photostationary state $\leq$ 0.03).[139]

Phytochrome phototransformations involve initial photoreactions followed by dark relaxation reactions, and several intermediates of phototransformation have been detected.[140]

Irradiating with natural light or high-intensity light of mixed R and FR, cycles phytochrome between Pr and Pfr with accumulation of some long-lived intermediates, especially a relatively weakly absorbing form, which directly precedes Pfr during transformation of Pr to Pfr. A precise determination of the in vivo proportion of intermediates cannot be made, but values in excess of 50% are possible. Though there is at present no evidence to support the idea that these intermediates are physiologically active, this possibility cannot be dismissed.

Besides the photochemical conversions, nonphotochemical reactions of phytochrome occur in vivo.[139] There is the *dark reversion of Pfr to Pr.* Since white light acts like R, at the end of the day, phytochrome is mainly in the Pfr form. Despite this, however, SDP and LDP become sensitive to R after several hours of darkness indicating that Pr is present in significant amounts. Thus, it was deduced that Pfr reverts spontaneously to Pr in darkness, although newly synthesized Pr can account for some portion of that found. Dark reversion was originally thought to be widespread, but is perhaps restricted to certain dicotyledonous tissues; it has not been detected in monocotyledons and Centrospermae. The thermochemical transformation of Pfr to a photochemically inactive form, known as *phytochrome decay* or *phytochrome destruction,* is widespread and dependent on metabolic activity. It is rapid under conditions which maintain a high level of Pfr (slow under conditions which maintain a low proportion of Pfr), and is thus a very important process in plants grown under fluorescent light which is very rich in R. New *synthesis of Pr* has been observed in several instances; it is detected in darkness. Unfortunately, we have little information on the relative rates of these reactions parameters which will be of prime importance in determining the kinetics of photoperiodic reactions. Thus, the model of phytochrome action including dark reversion, decay, and synthesis is:

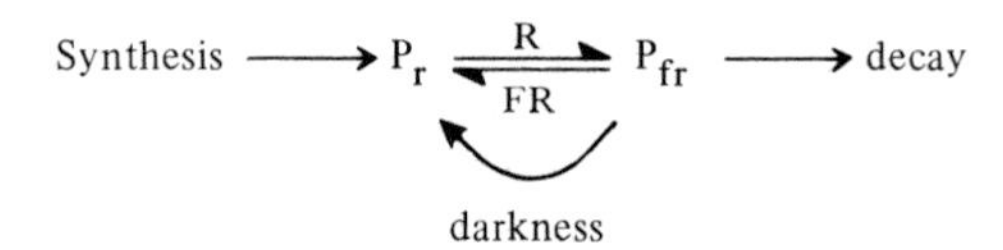

There is now increasing support for the hypothesis that phytochrome is associated with a membranal fraction and modulates one or more membrane activities.[139] Permanent association of a small fraction of phytochrome to particulates is disputable, but its association with particulate material following photoconversion to Pfr is consistently reported. Numerous studies in vitro or in vivo demonstrate indeed that irradiation with R increases phytochrome pelletability. Now the problem is to determine whether or not this is a measure of an *in situ* phytochrome-receptor interaction. According to Pratt,[139] there is as yet no definitive evidence which permits this conclusion; in fact, there is no evidence that suggests any receptor for Pfr. Further work is thus needed to solve the problems of site and mode of action of phytochrome.

C. Phytochrome and Photoperiodic Induction in SDP

1. Phytochrome Mediation of the Flowering Response during the Dark Period

In addition to *Xanthium,* soybean, and *Amaranthus,* the reversibility of the night-break effect by FR has been found in several SDP, including *Chrysanthemum,*[142] *Salvia occidentalis,*[118] *Pharbitis,*[143] and *Chenopodium rubrum.*[144]

The effectiveness of FR in reversing a R night break is influenced by the daytime light exposure. For instance, FR reversal is not observed in *Pharbitis* and in *Xanthium* when the photoperiod is very short (2 or 4 hr) or when the light intensity is low.[143,145] FR reversal in *Lemna paucicostata* 6746 is also dependent on light quality; it is demonstrated after a short photoperiod in blue, but not after R.[5]

In several SDP, the rate at which Pfr (produced by a R night break) reacts to inhibit flowering was estimated by inserting dark periods of various durations between the R

Table 2
FLOWERING RESPONSE[a] OF THE SDP *CHENOPODIUM RUBRUM* TO VARIOUS IRRADIATION TREATMENTS WITH BROAD BAND R AND BROAD BAND FR, STARTED 0.5 HR BEFORE THE MIDDLE OF 16-HR INDUCTIVE DARK PERIODS[b]

	Irradiation treatment			
Total duration of the treatments (min)	A Continuous FR	B 5-min R then FR	C 5-min R then dark until final 4-min FR	D R until final 4-min FR
10	8.3	8.0	8.2	7.9
20	7.5	7.3	7.2	7.3
30	6.8	6.3	6.7	6.5
40	5.9	5.7	5.5	5.4
50	3.0	2.9	3.1	2.7
60	1.0	0.9	1.1	1.2
70	0.3	0.3	0.4	0.3

[a] Stages of development of the inflorescence varied between the limits of 0 to 9.
[b] The plants were given five 16-hr dark periods.

and FR irradiations and determining the time taken to escape from reversibility by FR. Complete loss of FR repromotion occurs after an intervening dark period of about 30 min in *Xanthium* and 45 min in soybean.[137] In *Chenopodium rubrum* (Table 2, Column C) and *Chrysanthemum,* reversibility is maintained for longer periods: 70 and 90 min, respectively.[142,144] In *Chenopodium album,*[146] *Pharbitis,*[143] and *Kalanchoe,*[147] however, it is lost within 1 to 5 min. Thus, in these last species, when either the R exposure or the time elapsed between R and FR is too long, R-induced inhibition is only apparently irreversible.[148] In both *Xanthium* and Biloxi soybean, FR reversibility is lost very slowly when the temperature is lowered to 5°C during the interval between the R and the FR irradiation.[137]

The degree of reversibility of a R night break by FR diminishes when the time of exposure to FR increases (Table 2, Column B). In *Chenopodium rubrum,*[144] the reduction of the floral response is identical to that recorded when the plants are given either 5 min of R followed by darkness (Table 2, Column C) or extended R light (Table 2, Column D) for the same total period before 4 min of FR. Continuous FR exposures (Table 2, Column A) also result in a similar decline of inflorescence stages in plants that had no prior R treatment. *Chrysanthemum* behaves in the same way.[142] In *Xanthium,* however, the inhibitory effect of the prolonged FR irradiation is dependent on the length of the photoperiod: 60 min of FR fail to inhibit flowering until the photoperiod is reduced to 4 hr.[149]

Irradiation with R or FR establishes widely disparate photostationary states of 0.75 and 0.03; yet both completely prevent flowering. Clearly, inhibition of flowering is not proportional to the amount of Pfr, but results of maintaining for enough time the level of Pfr above some particular threshold. R is effective in relatively short exposure because it induces a high photostationary state, and when plants are returned to darkness, it takes a relatively long time for Pfr to decay below the threshold level. Relatively long exposure to FR is required because the proportion of Pfr is low, and on return to darkness, decay below the threshold value occurs relatively rapidly.

Maintenance of Pfr above the threshold probably accounts for the reduction of repromotion of flowering by FR noted with each successive R/FR cycle (Table 1).[137]

The effectiveness of cyclic lighting during the middle of an inductive dark period also indicates that the night-break reaction is a threshold type of response; in these treatments the dark intervals between the light pulses are not sufficiently long to permit Pfr to fall below the threshold level.

Since the night-break effect is very dependent on time of application during the dark period (Volume I, Chapter 3, Figure 3), low Pfr is not absolutely required throughout the night for photoperiodic induction of flowering in SDP. On the contrary, the requirement for a high Pfr level during the early hours of darkness is strongly indicated in several studies. For instance, flower initiation in seedlings of *Pharbitis* is inhibited by FR and repromoted by R at the beginning of a 16-hr inductive dark period. The action spectra for inhibition by FR and repromotion by R are shown by Nakayama and co-workers to be the same as those for other responses under the control of phytochrome.[148] Fredericq found that this response in *Pharbitis* is dependent on conditions during the photoperiod.[143] FR inhibits flower formation only when it is applied at the end of very short photoperiods of 2 to 4 hr; it inhibits also after 8-hr days if the intensity is low.

In *Chenopodium rubrum*,[150] *Lemna paucicostata* 6746,[151] *Xanthium*,[145] *Kalanchoe*,[147] and *Chrysanthemum*,[152] Pfr is also required during the first hours of the inductive dark period following very short photoperiods. In contrast, with longer photoperiods, so that the night is somewhat shorter than the critical, flower production is promoted by FR at the close of the daylight period in *Xanthium*,[136] *Pharbitis*,[143,153] and *Chenopodium rubrum*.[150,154] Results of Evans and King indicate that the inhibition of flower initiation by FR at the end of the light period is associated with the long dark, not the light period in *Pharbitis*.[153] They examined flowering as a function of the length of a dark period preceded by a 3-min exposure to either R or FR in seedlings. When the dark period exceeds 13 to 14 hr, FR inhibits and R promotes flowering; with shorter dark periods, flowering is promoted by FR.

The possibility that two Pfr processes are involved in the photoperiodic response of *Pharbitis* and *Chenopodium rubrum* is strongly suggested, using a physiological assay of phytochrome transformation developed by Cumming et al.[154] Brief exposure to various mixtures of R and FR at various times during a suboptimal dark period reveals that at each time there is a radiation mixture that is without effect on flowering; this is the *null response*. It is assumed that this response is obtained with the mixture which does not change the relative proportion of Pfr and Pr when night break begins. Thus a change in the relative amount of R giving a null response should reflect a corresponding change in the proportion of phytochrome in the Pfr form (resulting from dark reversion). Other mixtures of R and FR promote or inhibit. The radiation mixture giving *optimum flowering response* provides an estimate of the optimum proportion of phytochrome in the Pfr form in the course of an inductive dark period.

The null technique indicates that, in *Pharbitis*, reversion of phytochrome from the Pfr to Pr form does not occur until the 6th hr of darkness when a sudden change is observed (Figure 3A).[153] In *Chenopodium*, dark reversion occurs 2 hr earlier, i.e., around the 4th hr of darkness (Figure 3C).[155] In both species, plants exposed to a light rich in FR at the beginning of darkness to convert most of the phytochrome to Pr form, the proportion of R required for a null response increases progressively up to the time of dark reversion when it falls rapidly. Hence, an increase in the proportion

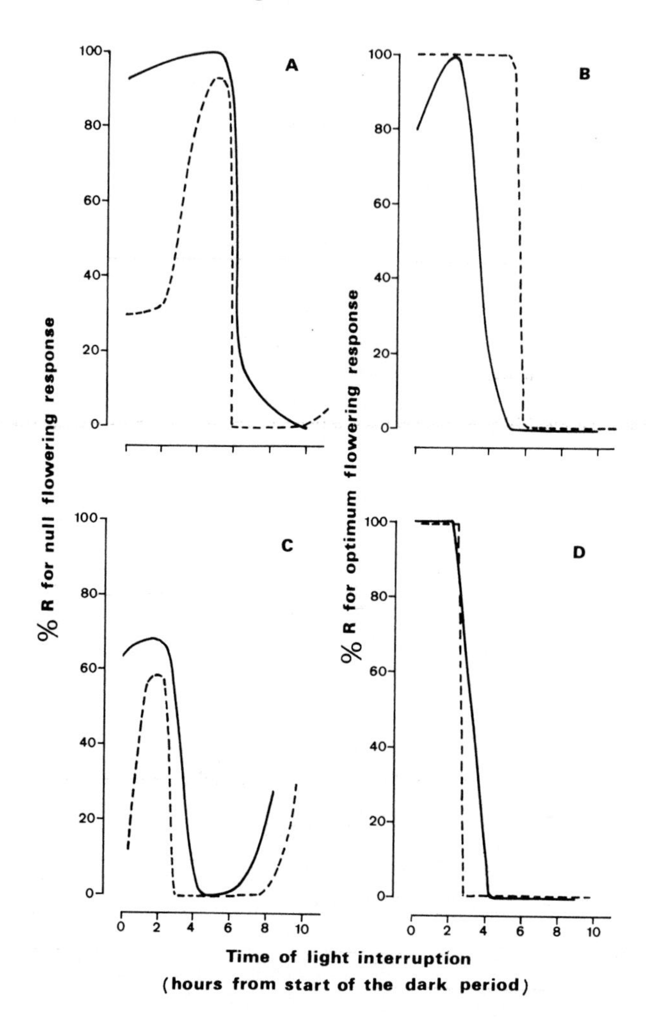

FIGURE 3. The percentage of R light in interruptions having (A, C) no effect, i.e., a null response, or (B, D) an optimum effect on flowering, as a function of time during a dark period. A, B: *Pharbitis nil.* The 14-hr dark period is preceded either by 3-min R (———) or by 3-min light rich in FR (– – –). (From Evans, L. T. and King, R. W., *Z. Pflanzenphysiol.*, 60, 277, 1969. With permission.) C, D: *Chenopodium rubrum.* A 9.5-hr dark period is preceded by 5-min R (———); a 10 to 10.75-hr dark period is preceded by 5-min light rich in FR (– – –). (From King, R. W. and Cumming, B. G., *Planta,* 108, 39, 1972. With permission.)

of Pfr, i.e., an apparent inverse reversion, is observed in darkness (Figure 3A and C). The radiation mixtures giving optimum flowering response indicate that during the early hours of darkness a high proportion of R is required in both *Pharbitis* (Figure 3B) and *Chenopodium* (Figure 3D). Thus, these results provide further support for the idea that photoperiodic induction of SDP implies a high Pfr process during the early hours of darkness followed by a low Pfr process (which proceeds only at very low Pfr values).

Direct measurements of Pfr transformation are required to test this inference. Unfortunately, because of the low level of phytochrome coupled with scattering and screening by chlorophyll, spectrophotometric assay of phytochrome in photoperiodically sensitive, green leaves is so far impossible. As demonstrated in recent studies with

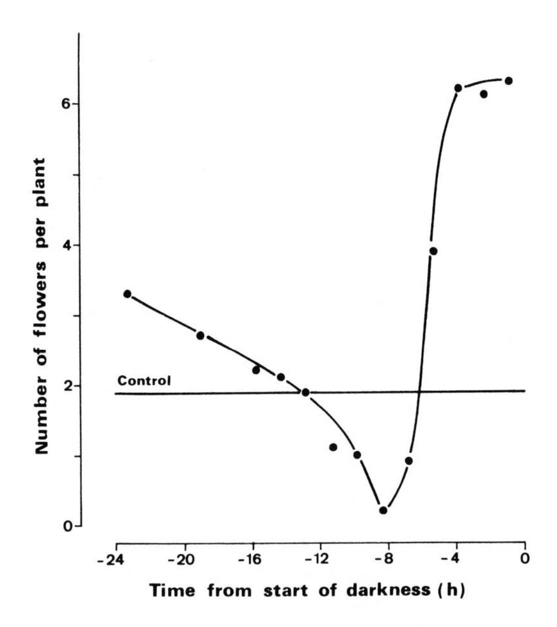

FIGURE 4. Flowering response of the SDP *Pharbitis nil* as a function of time of exposure to a 1.5-hr FR irradiation interrupting fluorescent white light prior to a 12.25-hr dark period. All treatments exposed to 5-min R immediately prior to darkness. Plot points are the midpoint of each interruption. (Adapted from King, R. W., *Aust. J. Plant Physiol.*, 1, 445, 1974.)

Pharbitis,[156] spectrophotometric measurements in partially green cotyledons are also inadequate for understanding the situation in green plants. There is much more rapid Pfr decay in partially greened tissues than in green cotyledons. However, comparison of spectrophotometric assays with the physiological assays of phytochrome transformation reveals rather good agreement between the two methods suggesting that the null technique reflects the true Pfr/P changes in tissues.

2. Phytochrome Mediation of the Flowering Response during the Day Period

Evidence is available that phytochrome mediates flowering response of SDP not only during the dark period, but also during the day period. In *Xanthium* and *Pharbitis,* the light quality during a light period, intervening between a subcritical dark period and an inductive dark period strongly affects flower formation. Action spectra indicate maximum promotion at 660 nm and inhibition at 730 nm.[157] In *Pharbitis* and *Chenopodium,* induced to flower by a single inductive dark period preceded and followed by continuous illumination, FR for 1 hr or longer, interrupting white light from fluorescent lamps or low intensity R, either inhibits or promotes flower initiation depending on the time of the interruption before the beginning of the dark period (Figure 4).[158] Evidence of R/FR reversibility indicates that phytochrome is the photoreceptor.

In dark-grown seedlings of *Pharbitis,* a prior exposure to light is required if a single inductive dark period is to be effective for flowering.[159] This requirement can be fulfilled by a single 24-hr irradiation or by as little as two 1-min R irradiations separated by a 24-hr dark period. The brief R irradiations are reversible with FR at very low energies: hence, it seems likely that phytochrome is active in the induction of sensitivity to the dark period.[160]

D. Phytochrome and Photoperiodic Induction in LDP

Can we say that phytochrome controls opposing reactions in LDP and SDP and

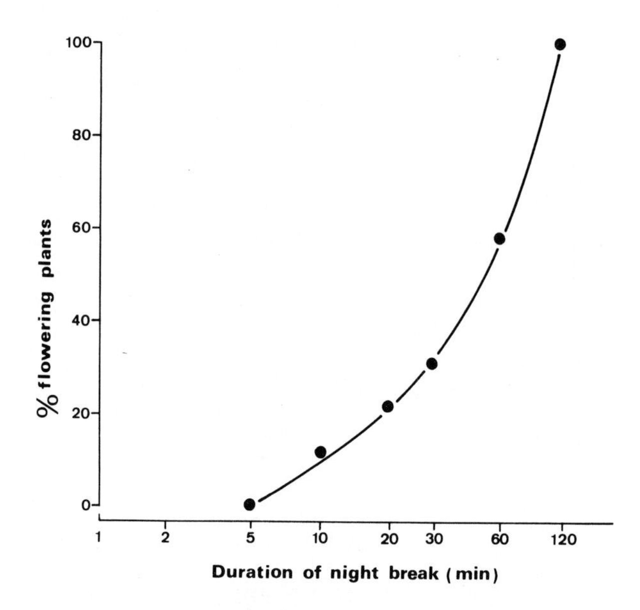

FIGURE 5. Flowering response of the LDP *Sinapis alba* as a function of length of a white night break given in the middle of six 16-hr dark periods. (From Hanke, J., Hartmann, K. M., and Mohr, H., *Planta,* 86, 235, 1969. With permission.)

everything else is the same? The situation is certainly not that simple. For example, the very brief interruptions of long nights, which inhibit flowering on many SDP, do not promote flowering in many LDP. Generally, much longer light exposures are required for LDP. Lane and associates found that 15-min fluorescent light breaks, given in the middle of a 16-hr dark period, fail to induce flower initiation in *Hyoscyamus,* beet, dill, or *Lolium* and are active in barley only after 5 weeks.[161] A 4-hr night break is effective. The need for a prolonged night break for flower induction in LDP is now well-documented (Figure 5).[58,78,162] Hence, the effect of the night break is not an all-or-none effect as it seems to be for SDP. Also in LDP higher energies are frequently needed. These results, only apparently, contrast with those obtained earlier by Borthwick and co-workers with barley and *Hyoscyamus* (Figure 1).[134,135] These studies were with plants in subthreshold daylength conditions where the energy required for an effective light break is greatly reduced compared to plants grown in 8- to 11-hr days.

R was shown to be the most effective when given for 1 or 2 hr as a night break in *Hyoscyamus,*[163] *Lolium* strain Ba 3081,[164] and *Lemna gibba* G3.[165] Light quality during the main light period frequently influences the response to the night interruption. In *Hyoscyamus,* for example, a 10-min R night break is effective in inducing flowering if the main light period is of predominantly blue light.[166] R interruption of the night is totally ineffective after photoperiods of green or R light. Similarly, a R night break does not induce flowering in *Lemna gibba* G3 after a main light period of R, but is effective after a white photoperiod.[165]

The action spectrum with prolonged night breaks is generally different from that recorded with short illuminations. In *Hyoscyamus,* sugar beet, and *Spinacia,* the peak of activity occurs at 719 nm with 8-hr night breaks.[167,168] A mixture of R and FR was shown to be more effective than R alone in *Sinapis* and *Dianthus,*[58,162] and similarly incandescent lamps with a high proportion of energy emitted in the FR region are more effective than fluorescent lamps,[161] which have little FR energy, for promoting flowering in various LDP. In fact, as demonstrated by Vince-Prue in *Petunia,*[55] it appears

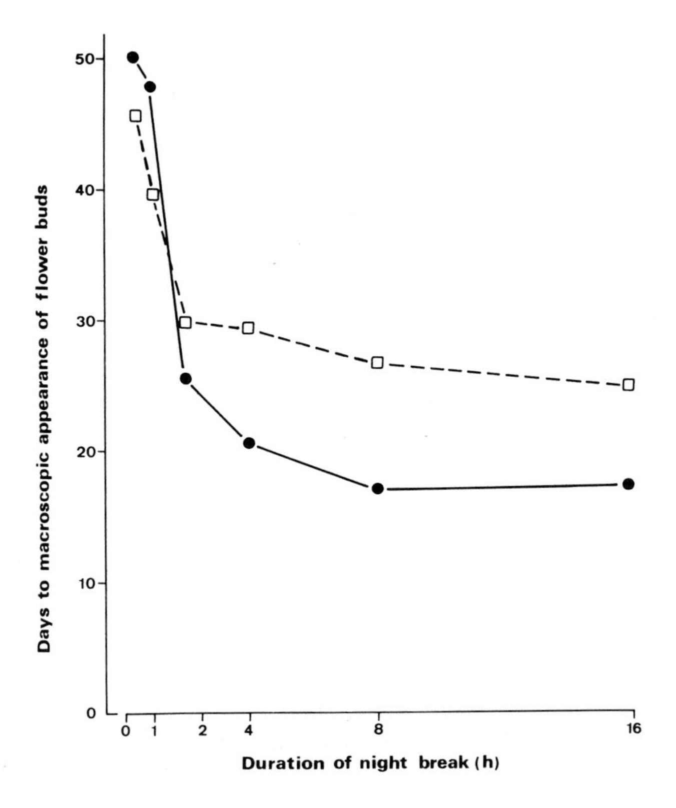

FIGURE 6. Flowering response of the LDP *Petunia* cv. Red Cascade as a function of length of R or R + FR night break given in the middle of a 16-hr dark period. SD control: 60.4 days to visible flower buds. (Copyright (C), 1975, McGraw-Hill Book Co. (UK) Ltd. From: Vince-Prue, D., *Photoperiodism in Plants,* Chap. 4. Reproduced by permission.)

that increasing the duration of the night break produces a change in the relative effect of R and R + FR (Figure 6).

Experiments with daylength extensions, rather than dark interruptions, also show that R alone or fluorescent light is generally not as effective as FR, a mixture of R and FR or incandescent light during the extension period.[57,161-164,169-172] The action spectrum shows a single peak in the FR region occurring between 700 and 720 nm as in wheat and *Brassica campestris,*[173] *Blitum capitatum,*[174] *Lolium,*[175] and *Anagallis* (Figure 7).[176] In *Blitum virgatum,*[174] the maximum occurs at 695 nm. Exceptions are *Fuchsia,*[177] which shows no marked acceleration of flowering when FR is added to a day extension with R and *Calamintha officinalis* ssp. *nepetoides,*[178] which reacts better to R than to FR.

In Crucifers, blue irradiation is most effective either as night interruption or as day extension as first shown by the pioneer works of Funke and Wassink and co-workers.[179,180] High responsiveness to blue is also observed in *Hyoscyamus.*[163]

Jacques and co-workers found that the action spectrum in the R-FR region is strongly dependent upon environmental parameters.[176,181] For instance, in *Anagallis,* inducing plants using high energies during the extension period, instead of low light intensity, shifts the region of greatest effectiveness towards shorter wavelengths (Figure 7).[176] In *Blitum virgatum,* R displaces FR in effectiveness as light intensity of the main light period is increased from 100 to 300 $W \cdot m^{-2}$.[181]

FR or incandescent light promotion of flowering in LDP changes during the course

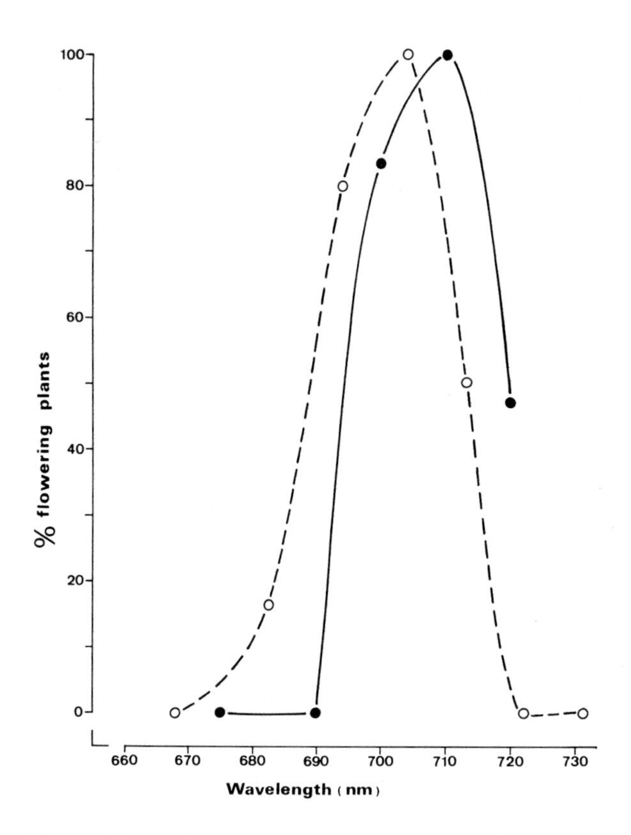

FIGURE 7. Action spectrum for photoperiod extension effect on floral initiation in the LDP *Anagallis arvensis*. (———) Plants exposed to 4 cycles each consisting of a 9-hr main light period followed by a 15-hr extension given by light from a spectrograph. The irradiance was 0.9 $W \cdot m^{-2}$ during the extension. (– – –) Plants exposed to a single cycle of 9-hr main light period followed by a 15-hr extension given by light from a spectrograph. The irradiance was 9 $W \cdot m^{-2}$ during the extension. (Adapted from Imhoff, C., Brulfert, J., and Jacques, R., *C. R. Acad. Sci.*, 273, 737, 1971.)

of a daily cycle as shown in *Hyoscyamus, Petunia,* dill,[161] *Dianthus caryophyllus,*[162] *Lemna gibba* G3,[165] and *Anagallis.*[182] In *Lolium* strain Ba 3081, a marked change in sensitivity to R and FR during an 8-hr extension of an 8-hr day in sunlight was also reported by Vince (Figure 8).[164] A 2-hr period of FR promotes flowering up to the 4th hr of a R extension and then inhibits, whereas R inhibits at the beginning of a FR extension and then promotes. The optimum time for R promotion of flowering is between the 9th and the 10th hr following the end of the main light period. These results are well-matched by observations on the Ceres strain (which is less sensitive to the R light break).[183] In this strain, peak effectiveness is in the FR region for irradiation just after the main light period while in the second part of the night, the most effective wavelengths are in the R region at 660 to 680 nm. Using the optimum flowering response technique (see Volume I, Chapter 4, Section I. C. 1), Evans determined when the shift occurs. As shown in Figure 9B, the percent of R required for optimum flowering response is almost 0 at the end of the daylight period, but it increases after a few hours of darkness.[183] Clearly, after the main light period, two processes are required in sequence for optimum flowering in LDP. The first process is favored by FR and the other by R. The null response technique indicates that reversion of the phyto-

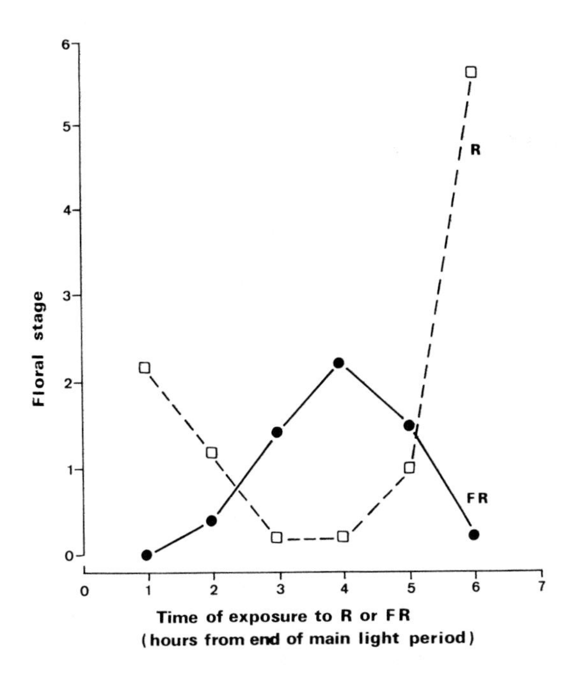

FIGURE 8. Flowering response of the LDP *Lolium temulentum* strain Ba 3081, as a function of time of exposure to a 2-hr R or FR interruption of a 7-hr extension in the alternate light, following 8 hr of sunlight. All treatments exposed to 1-hr R after the 7-hr extension. Ten LD are given. Data are plotted at the midpoint of the 2-hr interruptions. (From Vince, D., *Physiol. Plant.*, 18, 474, 1965. With permission.)

chrome to the Pr form seems to approach completion after 5 hr of darkness in *Lolium* (Figure 9A).

From the above considerations, it is difficult to decide whether phytochrome is the only photoreceptor involved in photoinduction in LDP. Wavelength dependence, the fact that the action spectrum in the R/FR region shifts towards longer wavelengths when established on a background of R,[175] and the opposite sensitivity to R and FR shortly after the end of the main light period in various LDP (Figure 8) clearly imply phytochrome. Furthermore, although R/FR reversibility is frequently impossible to show when long light exposure is required,[57,58] such a reversibility was demonstrated in some adequate experimental systems. In *Lolium,* Ceres strain, irradiation with R during the second half of a 16-hr long night causes flowering, and its effect was shown to be enhanced by exposure to FR during the first 6 hr after the main light period. During this initial period, brief exposure to FR is as effective as continuous irradiation with FR and is reversible by brief subsequent exposure to R.[183] Still in *Lolium,* but in the Ba 3081 strain, flowering is stimulated when a 9-hr day extension with R is interrupted in the middle with 2 hr of darkness. The effect of darkness is enhanced by FR at the beginning of the 2-hr dark period, and the effect of FR is nullified when followed by R.[184] In *Lemna gibba* G3, the R/FR reversibility was found just at the end of a 10-hr short main light period consisting of 3, 5, or 7 hr of R followed by 7, 5, or 3 hr of FR, respectively. In this case, the inhibitory effect of a 2-hr R break given during the first hour of the 14-hr long dark period is reversed by a subsequent exposure to FR for 15 min. The reaction is repeatedly reversible.[185]

Several hypotheses have been put forward to account for the promotive effect of FR and the requirement for prolonged exposure to light in LDP.[55] Especially, it was

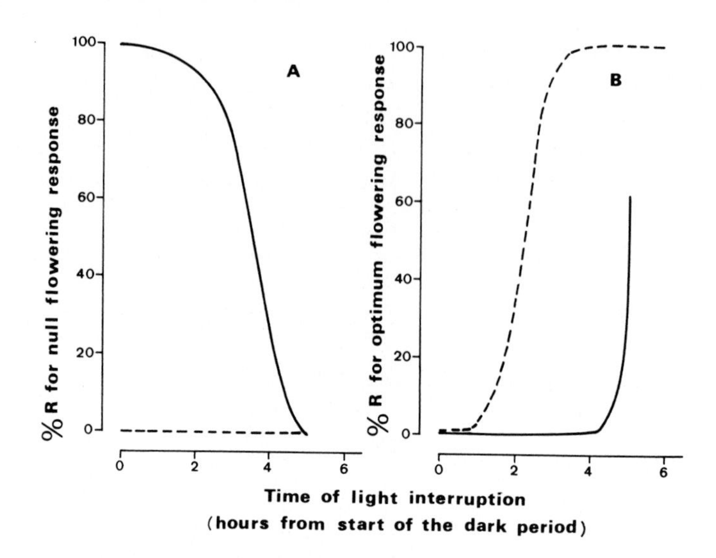

FIGURE 9. The percentage of R light in interruptions having (A) no effect, i.e., a null response, or (B) an optimum effect on flowering of *Lolium temulentum*, strain Ceres, as a function of time during a dark period of 6.5 hr following 8 hr of daylight and before 9.5 hr in R light. (———) Plants exposed to 3-min R at the start of the dark period. (– – –) Plants exposed to 3-min FR at the start of the dark period. (From Evans, L. T., *Aust. J. Plant Physiol.*, 3, 207, 1976. With permission.)

suggested that flowering in LDP is a type of "high irradiance response" (HIR).[186] Although it is now widely accepted that phytochrome is the photoreceptor in HIR, the mode of action of the pigment in mediating such response is controversial.[187] Jose and Vince-Prue believe that, in green plants and in light only, interconversions between different forms or associations of the pigment may be involved.[188] They suggest the term "dynamic" for this mode of action, in contrast to "static" for the R/FR reversible mode of action of phytochrome.

The existence of another pigment, perhaps a flavoprotein, has also been postulated to explain responses to blue light. However, it was frequently stressed that the different regions of maximum effectiveness to light may reflect different Pfr requirements and do not necessarily involve pigments other than phytochrome.[55]

CONCLUSIONS

Light perception in photoperiodism is essentially attributed to the pigment phytochrome (P) and, insofar as flowering is concerned, this is possibly the only photoreceptor. Phytochrome occurs as two interconvertible forms, one with an absorption maximum in the red portion of the visible spectrum (Pr), the other predominantly absorbing far-red radiation (Pfr). The absorption spectra of the two forms of phytochrome overlap considerably, and a total photochemical conversion is thus impossible regardless of wavelength used to irradiate the pigment. The proportion of Pfr to total phytochrome (Pfr/P) is critical in determining plant responses. In both SDP and LDP, it appears that two Pfr processes are involved: a low and a high Pfr process. The main difference between the two plant groups lies in the optimum timing for these reactions in relation to the main light period. In SDP, the high Pfr reaction is completed during the SD and the first hours of the long night; the low Pfr process occurs later in darkness. In LDP, the low Pfr reaction occurs first on transfer to dark; the high Pfr process

is required later. Since flowering may occur, under some circumstances, in continuous light as well as in continuous darkness in some SDP and LDP (see Volume I, Chapter 3), it seems that alternation in Pfr level in the daily cycle is not an absolute requirement. However, alternation is obviously superior for inducing flowering especially in SDP. In LDP, flowering is frequently most rapid under continuous irradiation with a suitable mixture of R and FR suggesting that at intermediate Pfr levels both the low Pfr and high Pfr processes can proceed.

II. TIME MEASUREMENT IN PHOTOPERIODISM

Many plants are sensitive to a difference in daylength of a few minutes,[21] indicating that they are able to measure time with great accuracy. Two different models have been proposed to approach the problem of photoperiodic time measurement. The first one views the clock as an *hourglass* which measures time by the accumulation or destruction of some unknown product; this model of the clock implicates a nonoscillatory timer. The second involves an endogenous oscillation, and the *circadian clock* has been implicated.

An essential problem was to determine whether photoperiodic induction of flowering involves only an hourglass timer, only a circadian oscillation, or some combination of both, since hourglass and oscillator are not necessarily mutually exclusive.[189] For several decades, results accumulated and available evidence supports the idea that both timers may play a role in the photoperiodic control of flowering.

A. The Hourglass Timer

The hypothesis that an hourglass timer is implicated in photoperiodic time measurement was formulated to account for results showing that in certain SDP the effective element in a daily cycle is the duration of the night rather than the duration of the light and that flower initiation occurs after exposure to a dark period longer than the critical independent of the duration of the associated light period as reported for *Xanthium* (see Volume I, Chapter 3). In *Pharbitis,* the number of flowers produced is proportional to the time in the dark beyond the critical length; furthermore, the quantity of flowers is very dependent on temperature.[190] In *Chenopodium rubrum,* though an endogenous rhythm is clearly involved in photoperiodic time measurement, it was demonstrated that flowering never occurs without satisfying a minimal requirement for 8 to 12 hr of darkness.[191] To interpret such results, it was suggested that a substrate decays over a period of time in dark until it falls below a threshold level. At this time, other reactions may start which eventually lead to flower formation. Light reinitiates the process, i.e., inverts the hourglass.

Considering only an hourglass model as the timer in photoperiodic induction of flowering raises some problems. Although in certain plants, e.g., the SDP *Pharbitis*[190] and the LDP *Hyoscyamus* (see Volume I, Chapter 3), the length of the critical night/day is strongly influenced by temperature suggesting the involvement of chemical reactions, in other species, including the SDP *Xanthium* and *Chenopodium rubrum,*[192,193] and the LDP *Sinapis* (Figure 10), the critical night/day length is little affected by temperature implying a physical process. Also the fact that the timing of the light interruption is generally affected neither by the length of the dark period nor by temperature suggests the involvement of a physical process though a temperature-compensated system of chemical reactions is also possible.

An hourglass timer might represent only one of various steps involved in an inductive night. This is suggested by data of Salisbury for *Xanthium* demonstrating that a low temperature (10°C) treatment given for 2 hr at the beginning of a dark period

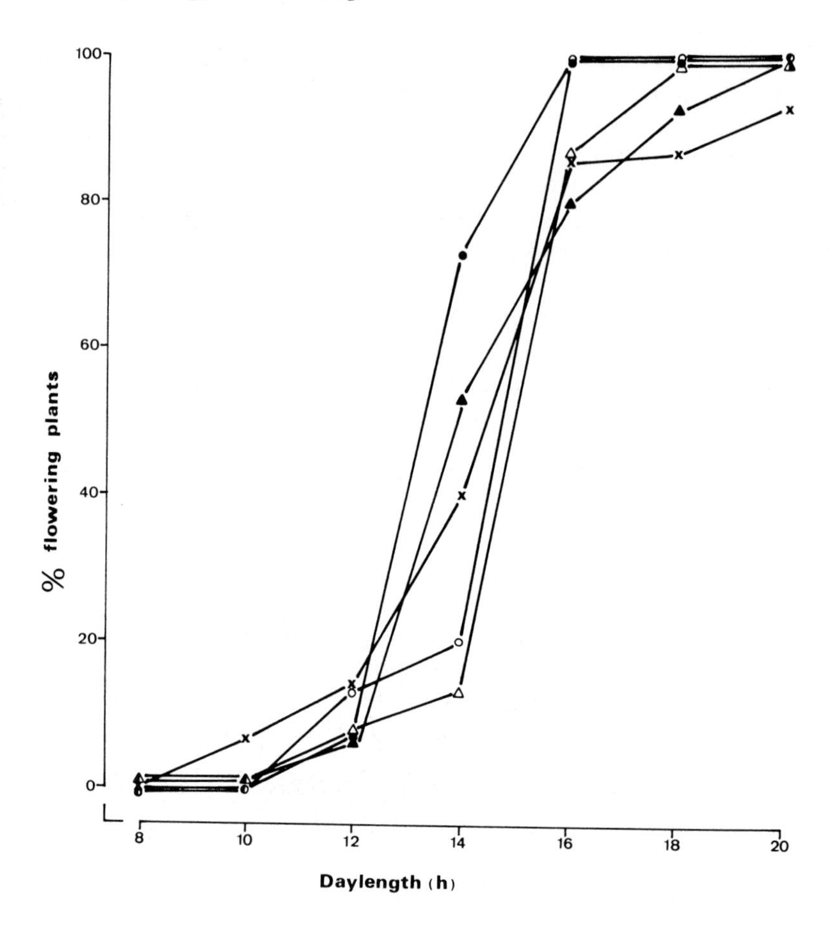

FIGURE 10. Flowering response of the LDP *Sinapis alba* to a single day of increasing length. Temperatures during the photoperiod were 15 (○), 20 (●), 25 (△), 30 (▲), and 36°C (X).

increases the critical night by about 45 min. When the low temperature period comes after the 5th hr, the flowering response is almost unaffected.[192]

B. The Circadian Clock

Bünning, in his hypothesis of the physiological clock, suggested that the measurement of photoperiod is executed by an endogenous free-running oscillation with a periodicity approximating 24 hr, i.e., circadian.[194-196] The oscillation was assumed to involve the regular alternation of two phases of about 12 hr. Bünning termed these 12-hr half-cycles the "photophile" phase and the "scotophile" phase and postulated that coincidence or noncoincidence of light with the scotophile phase was the factor determining the flowering response. Bünning's idea was that the photophile phase starts with the beginning of the light period; after 9 to 12 hr, the scotophile phase begins and extending the light period during this phase results in inhibition in SDP and promotion in LDP. Pittendrigh called this the "external coincidence" model, and stressed that it is essential to recognize the dual action of light which (1) acts as a photoperiodic inducer, (2) affects the entrainment mechanism of the oscillation, and thus controls the time of occurrence of the photoinducible phase.[189,197]

1. Experimental Evidence Implying Circadian Rhythms

There is a body of data, gained from three different experimental approaches, which

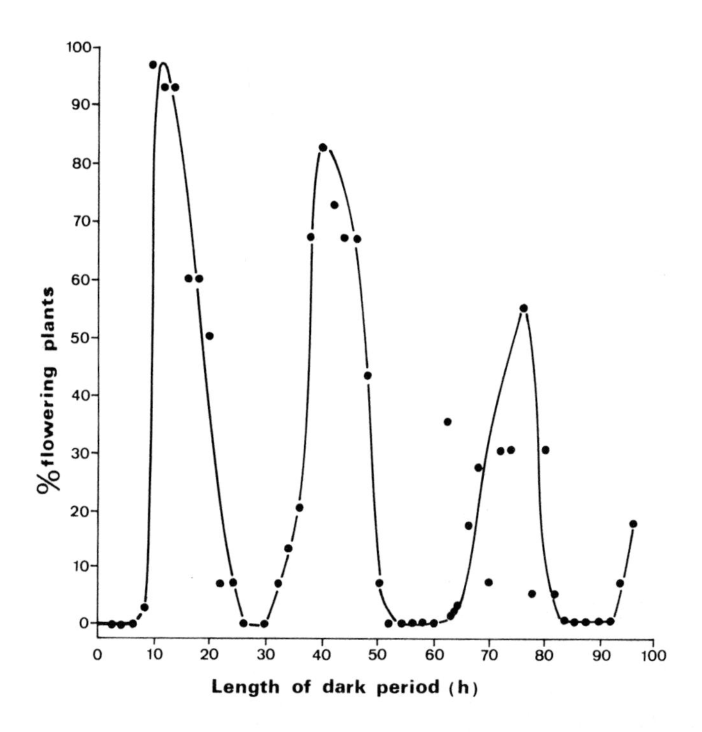

FIGURE 11. Flowering response of the SDP *Chenopodium rubrum* to a single long dark period of varied length interrupting continuous light. (Reproduced by permission of the National Research Council of Canada, from Cumming, B. G., Hendricks, S. B., and Borthwick, H. A., in the *Can. J. Bot.*, 43, 825, 1965.)

supports the view that a circadian rhythm of sensitivity to light controls flowering induction in photoperiodic plants.

a. Cycle Length (T Experiments)

By varying the length of both the light and the night period, it was found that cycle length is an essential factor in determining the flowering response of certain species. For example, with cycles up to 34 hr, Blaney and Hamner found that maximal flowering in Biloxi soybean occurs under 24-hr cycles provided the photoperiod is of 4 to 12 hr.[198] When an 8-hr photoperiod is used and the cycle length is varied up to 72 hr by varying the duration of the dark period, a rhythmic flowering response can be observed after seven of such cycles. Maximum flowering is recorded with 24-, 48-, or 72-hr cycles while a minimum response is produced by 36- and 60-hr cycles.[199]

In *Chenopodium rubrum,* a rhythm in flowering response is also displayed depending on the duration of a single dark period that interrupts continuous light (Figure 11).[154,191] Maximal flowering responses are produced by dark periods of 13, 44, and 71 hr indicating that the period of the oscillation is about 30 hr. In *Lemna paucicostata* 6746 the floral response, essentially the number of floral fronds, increases with increasing the length of a single dark period. Rhythmic variations in the response are nevertheless recorded. The time interval between the peaks approximates 24 hr suggesting the participation of an endogenous circadian rhythm.[200] A circadian periodicity in the flowering response of *L. paucicostata* to a single dark period of varied length was previously observed by Hillman.[201] In this experiment, the response was the result of an interaction between the dark period and seven repetitions of a skeleton schedule (see Section II. B. 1. c. in this chapter). The number of flowers per plant in *Kalanchoe*

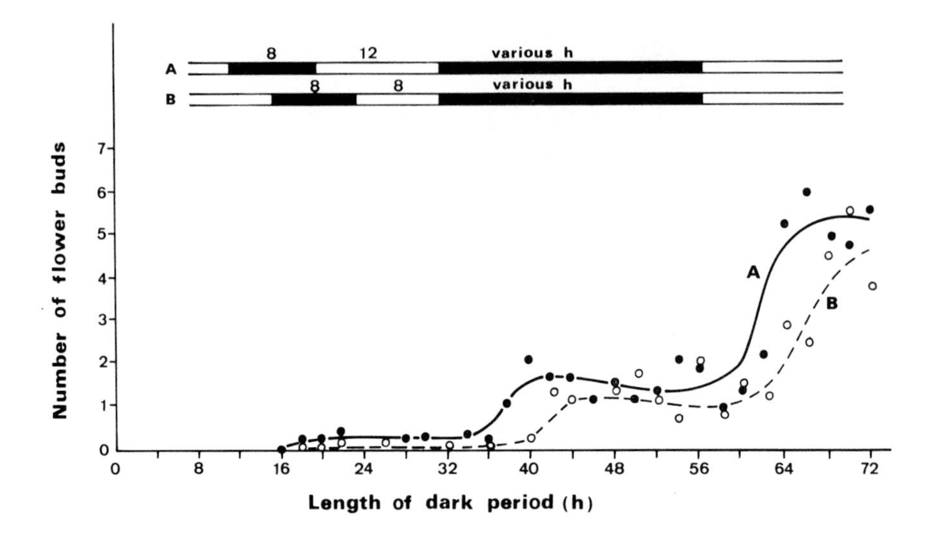

FIGURE 12. Flowering response of the SDP *Pharbitis nil* to a single dark period of various durations preceded by: (A) an 8-hr dark period followed by a 12-hr light period; (B) an 8-hr dark period followed by an 8-hr light period. Temperature was 18°C. (From Takimoto, A. and Hamner, K. C., *Plant Physiol.*, 39, 1024, 1964. With permission.)

also displays rhythmic variations depending on the duration of a single dark period preceding five inductive SD each consisting of a 9-hr photoperiod and a 15-hr night.[202]

We already reported that the flowering response of *Pharbitis* increases with the duration of a dark period interrupting continuous light suggesting that an hourglass clock is involved. A rhythmic response to night length can however be observed by introducing, prior to the inductive dark period, a photoperiodic treatment consisting of an 8-hr noninductive night period followed by either a 8- or a 12-hr light period.[190] With such a light-dark regime, increasing the duration of the dark period induces a stepwise increase in the flowering response with a periodicity of about 24 hr (Figure 12). This implies not only the hourglass timer, but also the circadian clock. An essential feature of Figure 12 is that the two curves are not superimposed when they are plotted in relation to the start of the inductive dark period, but only when plotted in relation to the beginning of the intervening light period. This suggests that the "light-on" signal of the light period initiates a circadian rhythm which affects the photoperiodic response in *Pharbitis.* This aspect will be discussed later.

In the LDP *Hyoscyamus,* a rhythmic flowering response to different cycle lengths has also been recorded in plants subjected to 42 cycles consisting of a 6-hr photoperiod (shorter than the critical daylength required for floral induction), followed by dark periods of various durations.[203] The maximal flowering response occurs with 12- , 36- , and 60-hr cycles; 24- , 48- , and 72-hr cycles result in reduced flowering response (Figure 13). The rhythm in the flowering response of *Hyoscyamus* thus exhibits peaks 12 hr out-of-phase from those of the SDP Biloxi soybean as required by the Bünning's theory.

One essential conclusion which can be drawn from the cycle experiments reported above is that photoperiodic timing in soybean, *Chenopodium rubrum,* and *Hyoscyamus* can not be simply a matter of measurement of either the dark or the light period. A similar situation probably exists in the LDP *Sinapis* which can be induced to flower when subjected to 34-hr cycles consisting of an 8-hr photoperiod, i.e., shorter than the critical daylength, followed by a 26-hr dark period,[63] but not in the SDP *Xanthium* where no flowering rhythm was found in response to a single long dark period.[204]

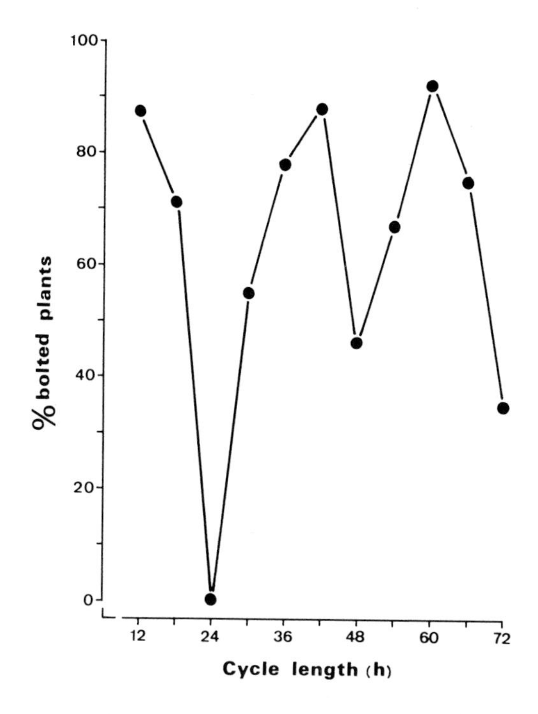

FIGURE 13. Flowering response of the LDP *Hyoscyamus niger* to long dark periods of various lengths. Plants were exposed for 42 days to treatments consisting of 6-hr photoperiods in conjunction with dark periods of various durations. (From Hsu, J. C. S. and Hamner, K. C., *Plant Physiol.*, 42, 725, 1967. With permission.)

b. Light Interruptions of Extended Dark Periods

By subjecting plants to long dark periods interrupted at different times by a light break, flowering of certain plants was shown to be a function of the time of the interruption. With 48-hr cycles initiated by a short light period of 6 to 10 hr, light interruptions of 30 min to 4 hr inhibit flower formation when given near the beginning or end of the dark period in the SDP *Kalanchoe,*[205] soybean,[206,207] and *Perilla.*[208] A light break near the middle of the dark period promotes flowering (Figure 14).

Essentially similar results were obtained with LDP, but in an opposite sense. Light interruptions near the beginning or end of the dark period promote flower formation whereas those near the middle inhibit or have no effect. This was observed in *Hyoscyamus* with 2-hr light breaks,[203,209] *Sinapis* using 8-hr light periods (Figure 14),[210] and *Lolium* with 4-hr light interruptions.[55]

In the case of *Sinapis,* it must be emphasized that only one *bidiurnal* cycle was used. Hence, floral initiation can be induced in this species by displacing the timing of a *single* 8-hr photoperiod. A so-called "displaced SD" treatment, consisting of delaying by 10 hr the timing of a single 8-hr SD (Figure 15), is advantageous in that induction in this LDP can be achieved without increasing the duration of illumination as compared to those plants in standard SD.

Usually, the two maxima for flower inhibition in SDP and flower promotion in LDP are approximately 24 hr apart, and the data can thus be interpreted according to Bünning's theory. However, Claes and Lang pointed out that these results also fit the alternative explanation that light interruptions act in conjunction with preceding or succeeding photoperiods.[209] In order to obviate this possible interaction between a light break and the nearest main light period, critical experiments have been devised using

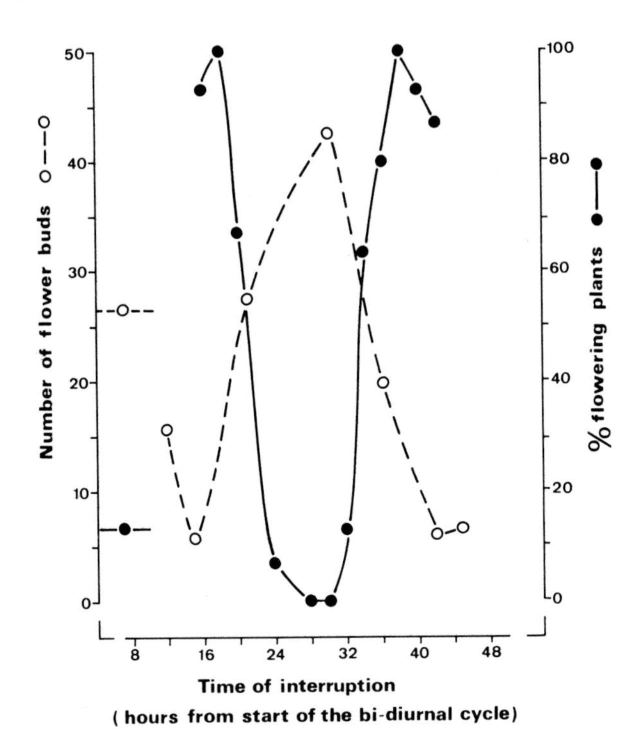

FIGURE 14. Flowering response of the SDP *Glycine max* (soybean) and the LDP *Sinapis alba* to a light interruption given at various times during the dark period of 48-hr cycles. For *Glycine* (O-O), the duration of the main photoperiod was 9 hr and the light interruption was 30-min long. The treatment was applied for 9 cycles. For *Sinapis* (●-●), the duration of the main photoperiod was 8 hr and the light interruption was 8-hr long. The treatment was applied for 1 cycle. Plot points are the midpoint of the time of exposure to the light interruption.[206,210]

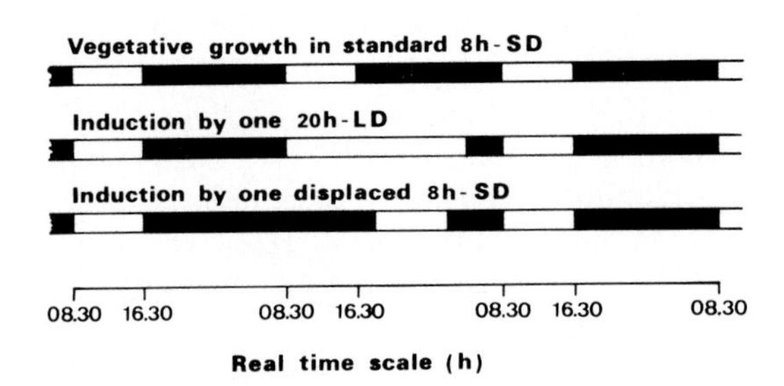

FIGURE 15. Schematic representation of different light/dark regimes used to control flower initiation in the LDP *Sinapis alba*.

72-hr cycles. In soybean, the 64-hr dark period following a main photoperiod of 8 hr is interrupted at various times by 4 hr of fluorescent light. Flower formation is inhibited not only by interruptions near the beginning and near the end of the dark period, but also by interruptions given near the 40th hr of the cycle; light breaks from 24 to 36 hr and from 48 to 60 hr promote flowering.[211] Similar results were obtained with other SDP such as *Kalanchoe* (Figure 16),[202,212] *Chenopodium rubrum,*[154] and *Lemna paucicostata* 6746.[200] In the LDP *Hyoscyamus*[203] and *Sinapis* (Figure 16),[63] peaks of

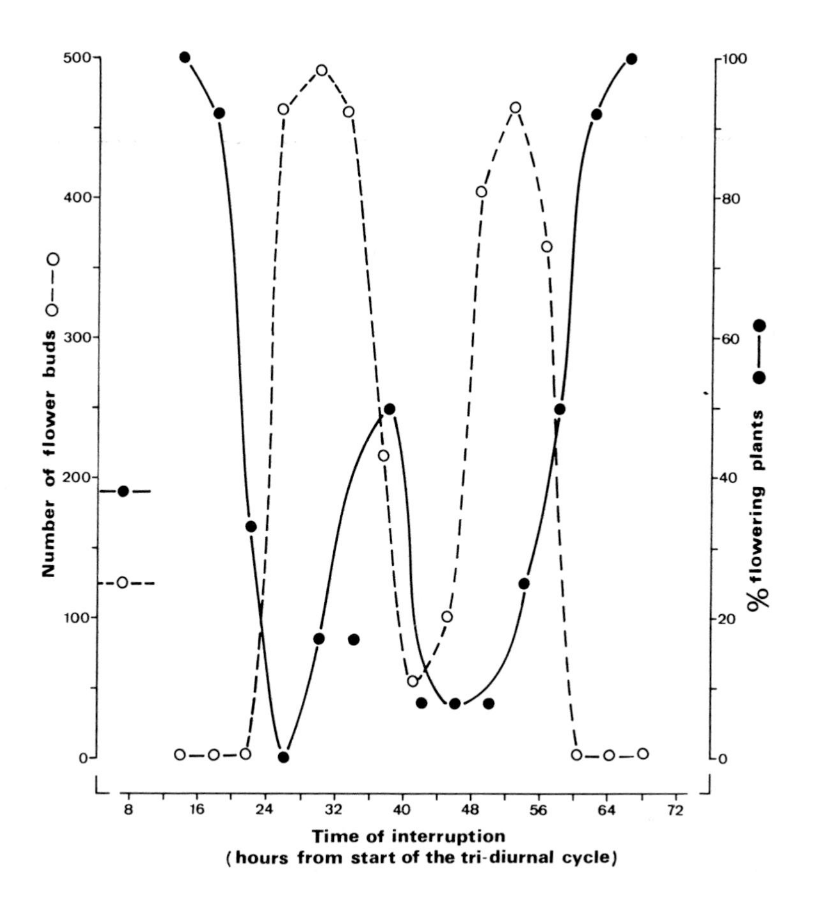

FIGURE 16. Flowering response of the SDP *Kalanchoe blossfeldiana* and the LDP *Sinapis alba* to a light interruption given at various times during the dark period of 72-hr cycles. For *Kalanchoe* (O-O), the duration of the main photoperiod was 10 hr and the light interruption was 1-hr long. The treatment was applied for 9 cycles. For *Sinapis* (●-●), the duration of the main photoperiod was 8 hr and the light interruption was 8-hr long. The treatment was applied for 1 cycle. Plot points are the midpoint of the time of exposure to the light interruption.[63,212]

promotion are observed at the times of maximal inhibition in SDP while the minima in flowering response coincide with maxima in SDP.

If 72-hr cycle experiments clearly imply the participation of an endogenous circadian rhythm in the control of induction in several photoperiodically sensitive plants, they also frequently offer support to the idea of Claes and Lang that light interruptions given near the beginning or the end of a long dark period interact with the nearest main light period. The two interpretations are not in fact mutually exclusive. A significant feature in experiments with 72-hr cycles is indeed that, in both SDP and LDP, the effect of the light breaks close to the start or the end of the extended dark period is greater than the effect of light interruptions given near the middle of the long night (see Figure 16). In some instances, as in *Anagallis,* the middle peak is even missing.[213] Interaction between the light interruptions given near the end of the long dark period and the following main light period is also strongly suggested by results of Shibata and Takimoto.[200] They found indeed that when the duration of the extended night is reduced to 60 hr, light interruptions given at the 48th hr inhibit flowering of the SDP *Lemna paucicostata,* while when given at the same time, in a 72-hr dark period, they promote flowering. However, if results are plotted in relation to the end of the dark period, it appears that in both cases, light interruptions given about 12 hr before the end of darkness inhibit flowering. Similarly, Wareing observed that light breaks given near the end of a 51-hr dark period inhibit flowering in soybean though they fall within a "photophile" phase.[206] Thus, a light break given near the end of the long dark period probably interacts with the following light period. The nature of this interaction is so far totally unknown. Finally, it must be stressed that a rhythmic flowering response to light perturbations of long dark periods has not been recorded in all species. In *Xanthium,* only one peak of inhibition is found at the 8th hr of the night,[204] while in *Pharbitis,* following a sharp peak of inhibition 8 hr after the start of the dark period, a second slight peak can be found about 24 hr later only when the temperature does not exceed 18.5°C.[190]

c. Skeleton Photoperiods

The action of photoperiods can be simulated by two short light pulses, separated by dark, one coincident with the start and the other with the end of the complete photoperiod. The pair of short pulses represents a "skeleton". For example, if 11(13) indicates a 24-hr cycle consisting of 11 hr of light and 13 hr of darkness, the cycle ¼ (10 ½) ¼ (13) can be considered as its skeleton. Many rhythms are entrained by skeleton photoperiods in the same way as by complete photoperiods: the pupal eclosion rhythm in *Drosophila pseudoobscura* provides a good illustration of this phenomenon.[189] The skeleton simulates the complete photoperiod up to 11(13), but the skeleton for 14(10) entrains the same rhythm as the skeleton for 10(14) with the only difference being a matter of phase.

Use of skeleton photoperiods in the control of the photoperiodic induction of flowering in *Lemna paucicostata* 6746 is instructive.[206,214] Cultures transferred out of continuous R light are subjected to seven 24-hr cycles with various complete or skeleton photoperiods defined by 15-min pulses (Figure 17). Flowering is promoted under complete photoperiods shorter than 10 hr, and no flowering is found when the light period exceeds 14 hr. Skeleton photoperiods up to 8 hr or longer than 14 hr are inductive, whereas skeleton photoperiods between 8(16) and 14(10) more or less inhibit flowering. A phase change occurs under skeleton photoperiods approximating 12 hr.

It must be noted that the skeleton schedule 11(13) permits much less flowering than its reverse schedule 13(11).[206] This indicates that the effect of the skeleton is dependent on the duration of the dark period separating the transfer from continuous light and the first light pulse. Figure 18 shows that by varying the length of this dark period, flowering response to the skeleton 13(11) exhibits a circadian periodicity. The rhythm apparently starts on transfer to darkness. Pittendrigh demonstrated that the behavior of *Lemna paucicostata* is consistent with the predictions of a theoretical model derived from studies of the entrainment by light pulses of the circadian oscillation in the pupa of *Drosophila pseudoobscura* and strongly supports the view that a circadian rhythm is involved in the photoperiodic control of flowering induction.[189]

In the LDP *Lemna gibba* G3, the critical photoperiod can be skeletonized since a 24-hr cycle functions as a LD when both the initial and the terminal portions of its subjective 12-hr day fraction are illuminated by 5-min pulses.[215,216] These critical initial and terminal portions were, respectively, termed L_1 and L_2 phases. A surprising feature is that no portion of the day other than the L_1 and L_2 phases are involved in induction of *Lemna gibba:* competence of these phases is exclusive.[215]

2. Phase Control by Light

Understanding how photoperiodic cycles effect the timing of a rhythm is of primary importance.[217] Phase control by the light-on signal was proposed by Bünning and found in *Pharbitis.* For instance, in seedlings subjected to a 12-hr inductive dark period at different times between 108 and 212 hr after the seeds were planted, fluctuations in

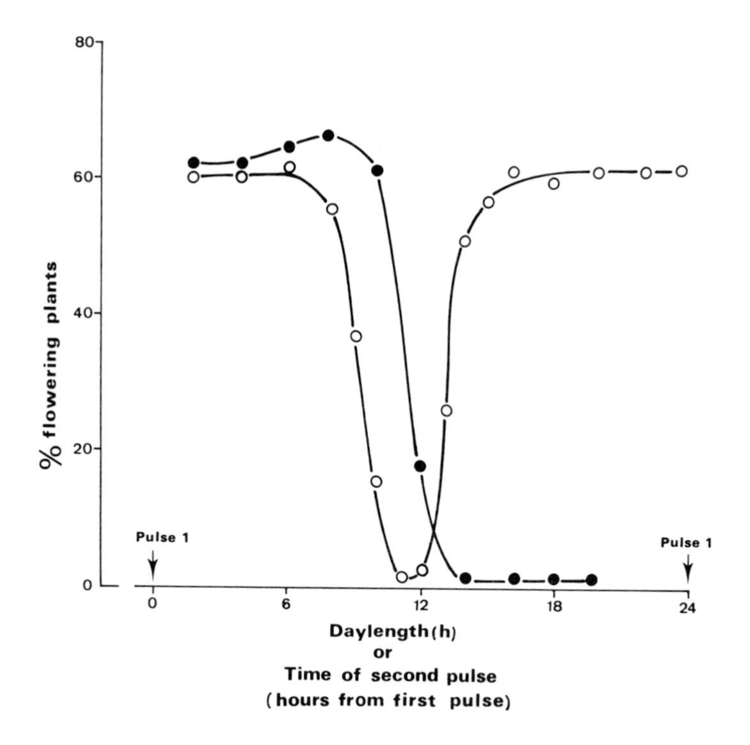

FIGURE 17. Flowering response of the SDP *Lemna paucicostata* 6746 to complete photoperiods or skeleton photoperiods defined by 15-min R pulses. Pulse 1 was given at time 0, 24, 48, ...hr and pulse 2 at various times (up to 24th hr) after pulse 1. The treatment was applied for 7 cycles. (From Oda, Y., *Plant Cell Physiol.*, 10, 399, 1969. With permission.)

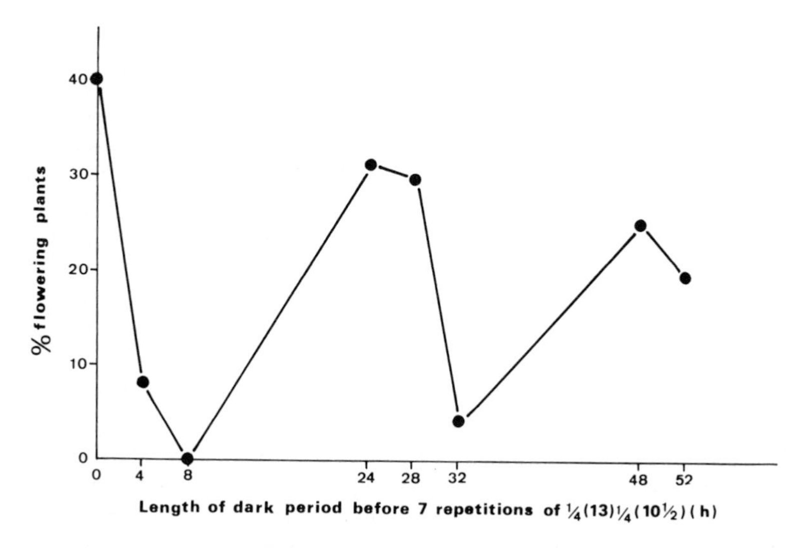

FIGURE 18. Flowering response of the SDP *Lemna paucicostata* 6746 to a single dark period of various lengths before seven repetitions of the schedule ¼ (13) ¼ (10 ½). R pulses were used. (Reprinted from *Am. Nat.*, 98, 323, 1964, by Hillman, W. S., by permission of The University of Chicago Press, Copyright 1964 by The University of Chicago.)

flowering response occur with a period approximating 24 hr.[218] This rhythm is probably initiated by the light the seedlings perceive when they first emerge from the soil.[219]

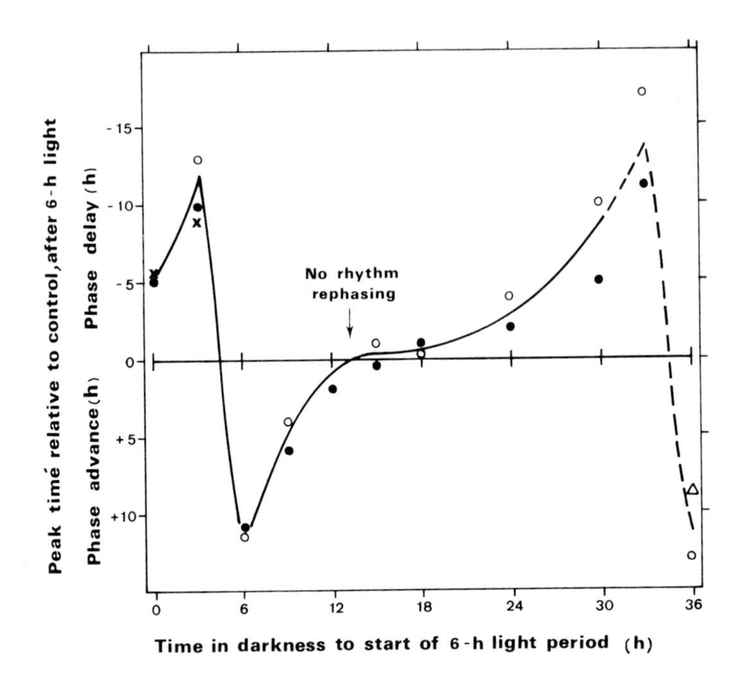

FIGURE 19. Phase response curve in the SDP *Chenopodium rubrum*. Phase delay or advance relative to the peak timing as affected by the time of exposure during darkness to a 6-hr light period. Rephasing of the first (×), second (●), third (○), or fourth (△) peak of the free-running rhythm. (From King, R. W. and Cumming, B. G., *Planta*, 103, 281, 1972. With permission.)

Takimoto and Hamner's work with 8- and 12-hr photoperiods (Figure 12) also indicates the occurrence of a light-on rhythm in *Pharbitis*.[160,190]

In numerous species a light-off signal starts the rhythm. This is apparent from Figures 11 and 18 which show that in *Chenopodium rubrum* and *Lemna paucicostata* a rhythm in flowering response is dependent on the duration of a dark period that follows continuous light. In *Pharbitis*, provided the temperature is dropped to 18.5°C, a light-off signal is also revealed by scanning a long dark period with 5-min R light breaks.[190] It was suggested that in this species two different oscillators are involved in the photoperiodic control of flower initiation: one is reset by light-on signals, the other by light-off. Induction occurs only when a particular phase relationship between both rhythms is established. This illustrates the "internal coincidence" hypothesis.[197]

In *Chenopodium rubrum*, King and Cumming demonstrated that both light-on and light-off signals influence the phase of the oscillation, but that their respective effectiveness is changed by photoperiod duration.[191] A 6-hr photoperiod given at various times, beginning from 0 to 36 hr in darkness, i.e., at different phases of the free-running oscillation of Figure 11, rephases the rhythm. Using rhythm peak times to indicate phasing of the oscillation, Figure 19 illustrates the relationship between the time of exposure to light and the resultant rephasing of the oscillation. Since (1) on transfer from continuous light to darkness the phase of the rhythm is controlled only by the light-off signal (see above) and (2) the period of the free-running oscillation is 30 hr (Figure 11), it is inferred that the light-off signal controls rephasing by a 6-hr light period starting at 0 and 30 hr. Therefore, the changing pattern of rephasing at other times can only be explained if both the light-on and light-off signals influence rhythm phasing. With 12-hr or more light periods, only the light-off signal controls the phase of the rhythm. Indeed, the peak of the rhythm always falls at a fixed time after the light-off signal (about 13 hr) (Figure 20).[191] Apparently, when the light pe-

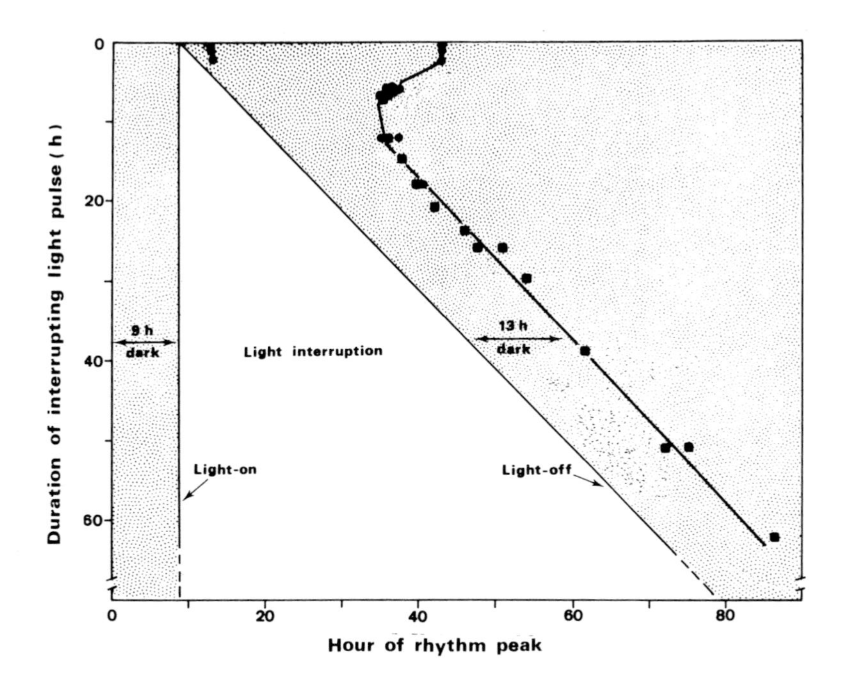

FIGURE 20. Time in darkness to maximum response at the first peak of the rhythm of flowering of the SDP *Chenopodium rubrum* given a single dark period interrupted with white fluorescent (■) or R (●) irradiation for various durations commencing at the 9th hr of darkness. (From King, R. W. and Cumming, B. G., *Planta*, 103, 281, 1972. With permission.)

riods exceed 12 hr, the rhythm is suspended and is reinitiated at a fixed phase in a subsequent dark period. The minimal duration of light required for rephasing of the rhythm lies between 2.5 and 6 hr. In *Lemna paucicostata* 6746 and *Pharbitis,* short light exposures are also unable to reset the clock.[201,220]

CONCLUSIONS

A diversity of mechanisms may be involved in the measurement of photoperiod. One model views the photoperiodic clock as an hourglass which measures time by the accumulation or the destruction of some unknown product. A second model implies a circadian oscillator. Since these two timers are not necessarily mutually exclusive, some combination of both may be involved in the photoperiodic control of flowering. The requirement for a dark period of a minimal duration in various SDP is suggestive of the operation of an hourglass timer. The idea that a circadian rhythm of sensitivity to light controls flowering in both SDP and LDP is supported by data of experiments in which plants are exposed to (1) cycles consisting of a fixed light period combined with dark periods of various durations, (2) light perturbations of extended dark periods, and (3) skeleton photoperiods. Most important for our comprehension of the control of flowering is the debate centering around the question of whether the endogenous oscillation is rephased by a light-on signal alone, a light-off signal alone, light-on and light-off signals together, or by two different rhythms controlled separately by a light-on and a light-off signal. In *Chenopodium rubrum,* the suggestion has been made that both light-on and light-off signals may control the phase of the rhythm, but as the duration of the photoperiod increases, there can be a change in their effectiveness for rephasing the rhythm.

III. PHYTOCHROME AND PHOTOPERIODIC TIME MEASUREMENT

A. Dark Reversion of Phytochrome and the Hourglass Timekeeping Model

Borthwick and Hendricks were the first to propose that the dark reversion of phytochrome could act as an hourglass timer.[221,222] They suggested that the inductive reactions cannot start before nearly all the Pfr has reverted to Pr. This hypothesis is supported by the observation that FR given at the beginning of a dark period reduces the critical night in various SDP (see Section I. C. 1 in this chapter). Obviously, as the extent of the reduction is generally limited to 1 or 2 hr, dark reversion of phytochrome occupies only a small part of the total inductive night.

According to King and Cumming,[155] the pattern of change in Pfr over the early hours of darkness in *Pharbitis* and *Chenopodium rubrum* as demonstrated by the null method (see Figure 3) also suggests that Pfr disappearance may play a role in photoperiodic time measurement. This idea is further supported by the fact that in *C. rubrum* single interruptions of darkness with R irradiations of 5 min lengthen the critical night only when the light break is given after the time of Pfr disappearance, i.e., after the 4th hr in darkness. Since short duration R does not rephase the rhythmic clock, [191] it is apparent that a timekeeper dependent on Pfr disappearance is involved in the photoperiodic control of flowering in *Chenopodium.* In *Pharbitis,* a delay in the time measuring process is also observed when a R light interruption is given after the time of Pfr disappearance.[153]

In both *C. rubrum* and *Pharbitis,* there is a 4-hr period between the completion of Pfr reversion and the length of the critical dark period. King and Cumming think that this correlation may be more than coincidental and gives significance to Pfr disappearance for time measurement.[155] It is evident however that such a timer is not involved in all photoperiodic species since dark reversion is not observed in Graminae and Centrospermae, even in species that are known to be photoperiodically sensitive.

B. Phytochrome and Circadian Rhythms

1. Phytochrome and the Inducible Phase

Action spectra for the effect of a light break given during the so-called scotophile phase strongly support the idea that phytochrome interacts with an inducible light-sensitive phase. However, some points remain unclear, and it is still difficult to explain how phytochrome acts in determining photoperiodic response. For instance, though a rhythmic response to R interruptions can generally be demonstrated during a long dark period in *Chenopodium rubrum,* soybean, and *Pharbitis,* a complementary rhythmic response to FR is generally not found. Usually, FR inhibits from the start of the dark period and the inhibition decreases almost linearly with delay of FR irradiation.[154,207,223,224]

Null response experiments in *Chenopodium* suggest that there is an endogenous rhythm controlling the transformation of phytochrome in darkness. The percent of R giving a null response is low from the 4th until the 8th hr in dark and increases thereafter indicating Pfr reappearance (Figure 3). Cumming and co-workers show that during a prolonged dark period, fluctuations in the percent of R giving the null response occur.[154] Minima are observed at the 30th and 59th hr. In the LDP *Lemna gibba* G3, the time at which there is a drop in the percent of R giving a null response during a 15-hr dark period, is not affected by temperature.[225] Furthermore, it changes rhythmically with diurnal periodicity depending on the length of the preceding light period.[226] This suggests that the time of phytochrome dark reversion is under the control of an endogenous circadian oscillation involved in the measurement of the preceding photoperiod.

In the LDP Wintex barley, Deitzer and associates found that addition of FR at

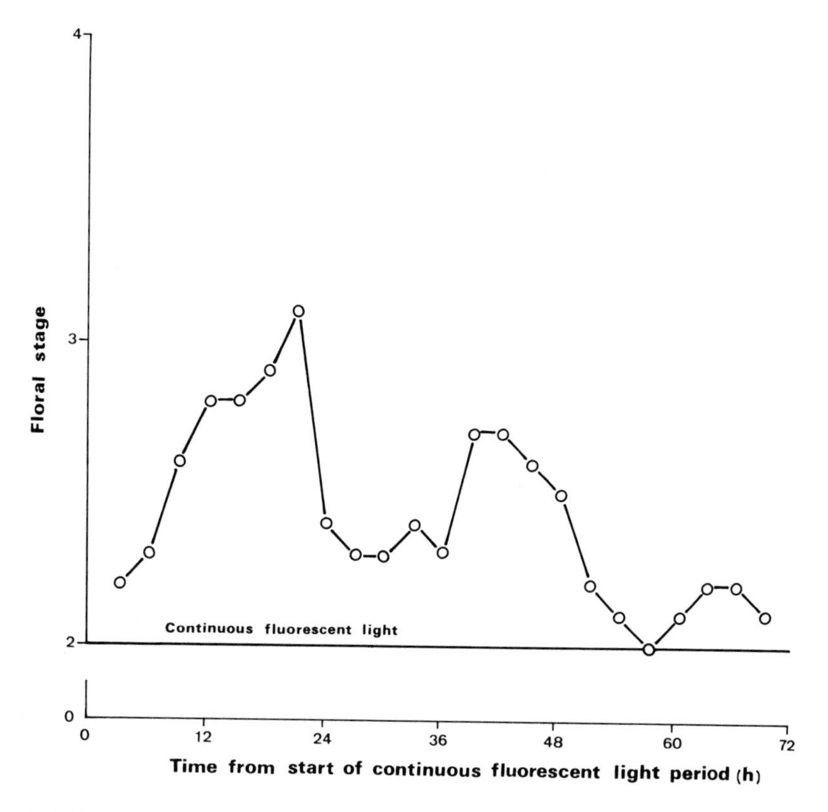

FIGURE 21. Flowering response of the LDP *Hordeum vulgare* (var. Wintex) to the addition of 6 hr of FR light at various times to a continuous 72-hr daylight fluorescent period which was inserted to interrupt 12-hr photoperiods. Plot points are the midpoint of the time of exposure to FR. (From Deitzer, G. F., Hayes, R., and Jabben, M., *Plant Physiol.*, 64, 1015, 1979. With permission.)

various times during a 3-day continuous fluorescent light treatment stimulates flower formation. The response shows a marked time dependence with maxima and minima recurring with circadian frequency (Figure 21), suggesting that there is a rhythm of sensitivity to a decreased Pfr/P ratio.[172] There is also a rhythm of changing sensitivity to the presence of Pfr in *Lolium*.[55]

2. Phytochrome and Rhythm Phasing

The involvement of phytochrome in rhythm phasing is suggested in some studies. In *Lemna paucicostata* 6746, the effect of a single variable dark period is still evident after seven daily repetitions of the skeleton schedule ¼ (13) ¼ (10 ½) (Figure 18) indicating that such schedules do not rephase the rhythm. Hillman shows that this is because the R light periods used in the skeleton are too short and that 4 to 6 hr of R light are required to reset.[5] Since six 15-min R light periods separated by 45-min dark periods are as effective as 4 hr of uninterrupted light, phytochrome is involved. Unfortunately, FR reversibility was not tested. In *Chenopodium rubrum*, a 6-hr light period which rephases the rhythm (Figure 19) can be partially replaced by 5-min R light pulses every 1.5 hr also suggesting the involvement of phytochrome in rephasing.[155] However, it is not possible to demonstrate clear FR reversibility. In dark-grown seedlings of *Pharbitis*, there is a changing response to the duration of the period of darkness given between the two short R irradiations required to induce sensitivity to the subsequent inductive dark period. This suggests that the first R irradiation may set the phase of an endogenous rhythm; since FR reversal is clearly established, phytochrome is strongly involved.[160]

C. General Remarks

Photoperiodic induction of flowering is dependent on events which occur when the plant is exposed to the inductive light schedule, but owing to the lack of criteria of floral induction in the leaf, the effect of these events is generally evaluated days or weeks later since we are forced to look at the production of flowers at the shoot apex for their estimation. Such a situation hinders understanding the precise function of phytochrome and circadian clock in the flowering process.[227-229] Between induction in the leaf and flower primordia differentiation, so many events are known to occur that the flowering process can stop at various steps before it can be measured. In order to obviate this problem, investigators of circadian timing in photoperiodic induction of flowering have utilized "overt" or "indicator" rhythms that can be readily observed.[228,229] Leaf and petal movements are generally used in these studies.[60,196,205,230-233] In *Lemna paucicostata,* CO_2 output has been chosen.[228,229]

With the search for processes that can be used as clock hands, a major question is to know whether or not the same timer is involved in the control of both the indicator and the flowering rhythms. Positive results were obtained with leaf movements in *Coleus* and soybean[60,231] and with petal movements in *Kalanchoe.*[202] In *Lemna paucicostata* 6746, CO_2 output is an accurate indicator of the timing process controlling photoperiodic induction of flower formation.[228,229] However, in several instances, the indicator rhythm does not correlate with the flowering rhythm. Examples are betacyanin accumulation in *Chenopodium,*[234] and leaf movements in *Xanthium* and *Pharbitis.*[230,232,233]

The leaves are the primary site of daylength perception, but the possibility that shoot tips are able to respond directly to changes in daylength cannot be dismissed (see Volume I, Chapter 3). It is known that the meristematic tissues contain the largest amount of phytochrome, and recently, Gressel and associates demonstrated that illuminating the plumule of *Pharbitis* with R partly suppresses flower initiation induced by a single 16-hr dark period provided R is given after the 8th hr in dark. Blue, green, and FR are ineffective.[235]

In *Chenopodium,* King suggested that flowering may be controlled by at least two distinct rhythms: one in the leaf; the other, possibly related to mitosis, in the apex.[236] In *Lemna gibba,* Oota and Tsudzuki also postulated the existence of a circadian oscillator in the meristem, distinct from another oscillator located in the nonmeristematic bulk of the frond.[216]

It thus appears that the meristem is probably not as passive as previously considered (see also Volume II, Chapter 1). The question of how timing processes in the meristem modulate its response to leaf-generated floral stimuli is of major concern.

CONCLUSIONS

There are various possible interactions of phytochrome with the photoperiodic clock. When an hourglass clock operates, dark reversion of phytochrome could generate the timer or one of the processes involved in this clock. When the circadian clock is the timer, there are two possible ways in which phytochrome may be involved. First, the pigment could interact with an inducible light-sensitive phase and act as a signal inducing or inhibiting flower initiation. Second, phytochrome may function as a synchronizer and affect the phase of the rhythm(s) involved in the measurement of photoperiod.

Irrespective of the nature of the timer, transitions from light to dark must be per-

ceived, and phytochrome is a candidate for such a function. Pigment cycling (see Section I. B. in this chapter) which occurs in light and stops on transfer to darkness could provide the on-off signal.[237]

The lack of an assay for induction in the leaf hinders comprehension of the precise function of phytochrome and circadian clock in the flowering process. Also, the possibility that shoot tips respond directly to photoperiod raises the question of how timing processes in the meristem modulate its response to leaf-generated floral stimuli.

Chapter 5

CONTROL BY LOW TEMPERATURE

TABLE OF CONTENTS

I. INTRODUCTION

The utmost importance of low temperatures, i.e., temperatures below those optimal for growth, in the control of flower initiation in some plants was known in the 19th century, and then definitively established by Gassner in 1918. A story of the discovery of "vernalization", a word which is an English translation of the Russian word "Yarovisation", can be found elsewhere.[24,238]

Strictly speaking, vernalization should be used only for the induction or promotion of flowering by a low-temperature treatment.[112] As a rule initiation of flower primordia does not occur at vernalizing temperatures, but only occurs after plants are moved to higher temperatures more favorable for growth. Thus, the action of chilling is inductive similar to that of daylength, hence, the word "thermoinduction". However, in some Cruciferae, e.g., *Brassica oleracea gemmifera* (Brussels sprouts), *B. rapa* (turnip), *Matthiola incana* (stock), and in bulbous irises, flower initiation may occur only at low temperatures.[21,239] Whether one should make a clear distinction between this *direct* effect and the more general *inductive* effect of low temperatures is not at all clear. In many plants exhibiting usually the inductive type of response, initiation of flowers may start during the cold treatment itself (direct response) if the latter is sufficiently long. Examples of this are the biennials *Lunaria annua,*[104] *Campanula medium,*[113] *Cheiranthus allionii,*[240] *Oenothera biennis,*[241] and *Daucus carota* (carrot),[242] and the perennials *Geum urbanum,*[112] *Cynosurus cristatus, Dactylis glomerata,*[243] and *Poa pratensis.*[244] Thus, these plants exhibit one or the other type of response depending on the duration of the chilling treatment, suggesting that low temperature action is basically the same in both cases. Accordingly all the promotive effects of cold on flower initiation will be considered here as vernalizing effects.

In cold-requiring plants, as in photoperiodic plants, there are species with an absolute (or *qualitative*) and species with a facultative (or *quantitative*) cold requirement. As a rule, plants with a facultative requirement can be vernalized as imbibed seeds, whereas those with an obligate requirement cannot and must reach a certain size for attainment of responsiveness to cold. Grossly speaking, biennial and perennial plants have an obligate requirement while winter annuals only a facultative one, but exceptions to this generalization are known. As pointed out by Lang,[21] the distinction between these two kinds of requirement is, as in photoperiodism, most probably one of degree. Indeed, closely related species or even races within a species may differ in their vernalization responses, ranging from great sensitivity to virtual indifference (Figure 1). Also, after crossing the annual strain of *Hyoscyamus,* which has no cold requirement at all, with the biennial strain, which has an obligate cold requirement, plants of the F_1 generation possess a facultative cold requirement.[245]

II. DURATIONS OF TREATMENT AND EFFECTIVE TEMPERATURES

The length of the chilling treatment and the range of temperatures that are effective vary with species and even with variety. Also, the effectiveness of a certain exposure to low temperatures is markedly dependent upon the age of plants, a topic that will be discussed in Volume I, Chapter 7. Usually a 1- to 3-month exposure to low temperature is required for effective thermoinduction in winter annuals, biennials, and perennials. The fact that vernalization is such a slow process probably explains why the number of fundamental studies devoted to it is far less than to photoperiodic induction of flowering. Chilling treatments of a few days, up to 2 weeks, may however have a promotive effect in plants with low vernalization requirements. Extreme cases are those of stock, *Apium graveolens* (celery) and *Anthriscus cerefolius* (chervil) in which a sig-

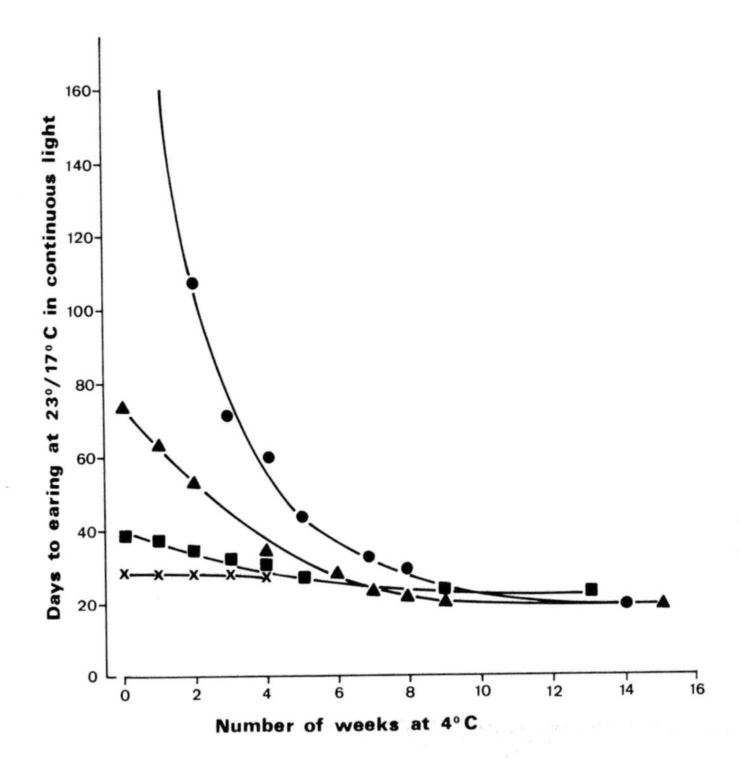

FIGURE 1. Responses by several ryegrass species to vernalization in SD at 4°C. After chilling, plants are grown in continuous light at 23/17°C (day/night). ×, *Lolium temulentum;* ■, *L. multiflorum;* ▲, *L. multiflorum* × *L. perenne;* and ●, *L. perenne.* (From Evans, L. T., The influence of temperature on flowering in species of *Lolium* and in *Poa pratensis, J. Agric. Sci.,* 54, 410, 1960. Reproduced by permission of Cambridge University Press.)

nificant promotion of flowering has been observed following a 1- or 2-day low-temperature treatment.[246-248]

Owing to the use of a protracted cold treatment the polycarpic plant, *Geum urbanum,* has become a monocarpic plant in the hands of Tran Thanh Van-Le Kiem Ngoc.[31] In this species, the terminal meristem of the rosette as well as some old axillary meristems never flower in response to a chilling treatment of "normal" duration, i.e., 2 to 3 months, and thus serve to continue vegetative growth year after year. All these insensitive meristems can be brought into flowering by an "unusual", 1-year-long cold treatment. The final result is that *Geum* dies after its first flowering and thus is transformed into a monocarpic plant. This kind of experiment is difficult to extend to other species since cold treatments of extremely long duration often result in a deterioration of the plants.

Thermoinduction is clearly a quantitative process, i.e., the longer the chilling treatment the more effective it is up to a certain maximal value (Figure 1). Using a system of floral stages to estimate the progress of thermoinduction in Petkus winter rye after different durations of cold treatment, Purvis obtained the sigmoid curve of Figure 2 (curve A), which can be fitted to the equation of an autocatalytic reaction.[249,250] Caution must be exercised in drawing conclusions about the kinetics of thermoinduction from such an analysis in which morphological change only is the dependent variable (see Volume I, Chapter 1). Indeed, with another measuring system, the form of the vernalization response curve may be different.[251]

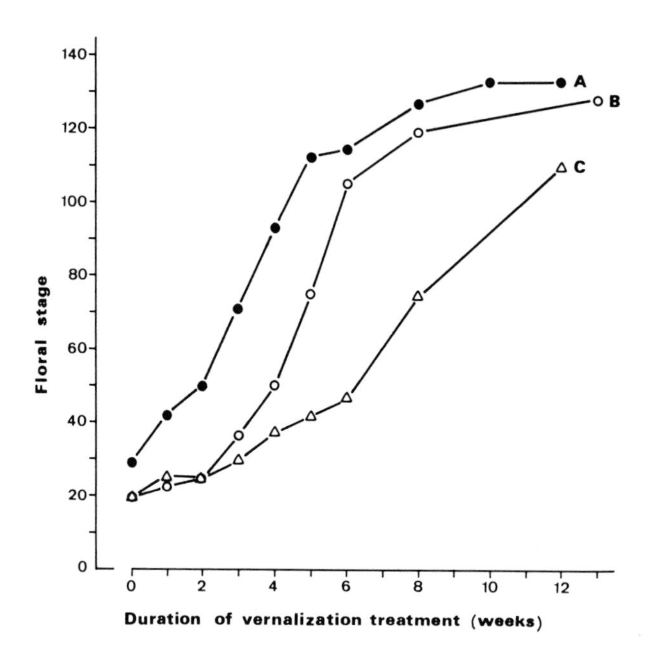

FIGURE 2. Progress of vernalization in *Secale cereale* var. Petkus (winter rye) with duration of treatment. A: intact grains; B: excised embryos provided with 2% sugar; and C: excised embryos without added carbohydrate. Flowering is measured by a system of floral stages. (From Purvis, O. N., *Encyclopedia of Plant Physiology*, Vol. 16, Ruhland, W., Ed., Springer-Verlag, Berlin, 1961, 76. With permission.)

Temperatures in the +1 to +7°C range are usually most effective for thermoinduction.[21,112,250] However, temperatures below 0°C and as low as −6°C, have been found effective in certain cereals. Other plants from warmer regions, such as *Olea europaea* (olive), may have higher ranges with optima at +10° to +13°C.[252] This makes it clear that vernalization may well take place during a period of concurrent, but reduced vegetative growth.

In the best investigated cases it appeared that the most successful temperature is dependent on the duration of the treatment. As a rule the optimum temperature decreases when duration increases (Table 1). From this, Lang concluded that any temperature in the effective range results finally in the same, maximal thermoinduction provided that the duration of the treatment is sufficiently extended.[21]

In some cases, e.g., celery, *Lolium perenne, Chrysanthemum,* and *Geum,* effective vernalization can take place even when the plants are exposed to moderate nonvernalizing or less vernalizing temperatures for part of each day, at least when the highest temperature coincides with the daily light period.[31,247,251,253] In other plants, e.g., winter rye, *Cheiranthus allionii,* and cabbage, such "interruptions" of the cold treatment greatly reduce or nullify its effect.[240,254,255] The complexity of the temperature requirement in some species is well-illustrated by the case of *Oenothera biennis.* This plant will not flower after continuous exposure to temperatures of 11° or 3°C, but absolutely requires a discontinuous cold treatment with daily alternations of 11° and 3°C. However, periods of 15° or 18°C completely abolish the effect of a preceding vernalizing treatment.[241]

III. OTHER PREREQUISITES FOR EFFECTIVE SEED VERNALIZATION

Presence of oxygen is apparently required for seed vernalization since chilling in a nitrogen atmosphere is ineffective in winter cereals and radish.[21,250]

Table 1
OPTIMUM TEMPERATURES (IN °C) FOR VERNALIZATION IN RELATION TO THE DURATION OF THE CHILLING TREATMENT

Species	Duration of vernalization (weeks) 2	3	4	5	6
Petkus winter rye	+7	+5	+1	+1	+1
Biennial *Hyoscyamus*	+10		+6		+6 to +3
Arabidopsis race *St*	+4				+2
Lactuca sativa cv. Imperial 456			+2		+0.5

From Napp-Zinn, K., *Temperature and Life,* Precht, H., Christophersen, J., Hensel, H., and Larcher, W., Eds., Springer-Verlag, Berlin, 1973, 171. With permission.

Dry seeds cannot be vernalized, but imbibed seeds are sensitive. About 40 g of water per 100 g of air-dried caryopses is often sufficient for seed vernalization, although not for germination, of cereals.[250,256] In seeds containing fat and/or protein as reserve materials, up to 100% on a dry weight basis may be necessary.[256]

IV. THE PERCEPTION OF LOW TEMPERATURE

Curtis and Chang demonstrated that chilling is perceived in celery at the crown by the shoot apical meristem and/or the surrounding young leaves and not by the rest of the plant.[257] A similar situation is common in other biennials and perennials. Thus in beet, *Chrysanthemum,* and olive, exposure of plant parts, other than the shoot tip, to low temperature does not result in vernalization.[258-260] Successful reciprocal graft transfers of shoot tips between vernalized and unvernalized *Hyoscyamus* and *Althaea rosea,* and successful vernalization of excised shoot tips of carrot and cabbage clearly demonstrate that the shoot apex is the perceptive tissue.[21,261,262] In winter cereals the embryo alone needs to receive cold treatment; even parts of the embryo are susceptible to treatment as long as the stem apex is included.[263] Several of these studies have further demonstrated that there is a requirement for sugar in the medium in which the excised shoot apices or embryos are vernalized (Figure 2, curve B).[262,264] In the absence of the carbohydrate substrate, vernalization is slow, but is ultimately accomplished (Figure 2, curve C), probably at the expense of the reserves of the embryo. These observations have led to the conclusion that plant parts, such as mature leaves, endosperm, and other storage organs whose removal may abolish the sensitivity to cold,[31,240,241,260,265] play only a trophic role during the chilling treatment in supplying raw materials to shoot apices.[21,265]

This view prevailed until it was disclosed that such different plant parts as the leaf or petiole cuttings of *Lunaria annua* or root segments of *Cichorium intybus* (chicory) regenerate flowering shoots after an appropriate cold treatment.[104,266-268] By removing the basal ½ cm of the petiole of leaf cuttings of *Lunaria* immediately after the cold treatment, Wellensiek found that shoots that regenerate are now vegetative instead of reproductive.[266] This suggests that the perception of cold by the cuttings is localized in this region of the petiole, which is also the site of shoot regeneration and thus of mitotic activity. Leaves on intact *Lunaria* plants can also be vernalized, provided they are not fully expanded at the start of the cold treatment. The younger the leaf, presumably the greater the mitotic activity, and the more effective is vernalization.[266] Similarly, Pierik showed that cotyledon cuttings or petiole segments of *Lunaria* are hardly vernalized, even after prolonged exposures to low temperature, if entirely devoid of regenerated buds.[104] In chervil, the most sensitive stage to a chilling treatment occurs at the onset of mitoses in the germinating embryo.[248] Wellensiek concluded from all of this that vernalization only takes place when dividing cells are present during the cold treatment.[266] This would explain why shoot apices are the main site of perception for low temperature. This concept, however, has not gone unchallenged. Successful vernalization has been obtained under conditions of temperature which apparently excluded mitotic activity.[85,240,269] We cannot be sure however that in these cases there is not a slow progress of cells along the mitotic cycle. Such activity is by no means excluded on *a priori* grounds, and this possibility can be checked by a [^{3}H]thymidine labeling of appropriate duration. As this has not yet been done with these materials, there is no real basis to reject Wellensiek's concept. Recall here that cell division as well as organ formation and development can proceed in some plants under snow cover at 0°C.[270]

V. INTERACTIONS OF VERNALIZATION AND OTHER ENVIRONMENTAL FACTORS

The effect of a cold treatment may be influenced by other environmental parameters, and these interactions are certainly not less complex than those described in Volume I, Chapter 3 in the field of photoperiodism.

A. High Temperature

The best studied of such interactions, that between chilling and subsequent heat treatments, reveals at least three situations:

1. High temperatures given for a few days may, under some circumstances, annul the effect of a previous cold treatment, a phenomenon called "devernalization" (Figure 3). This observation first made in Petkus winter rye by Gregory and Purvis[270a] was later extended to many other cold-requiring plants, e.g., other winter cereals, sugarbeet, biennial *Hyoscyamus*, celery, *Dianthus barbatus, Cheiranthus allionii, Oenothera biennis*, and *Arabidopsis*.[21,240,241,271]
2. High temperatures alone have no devernalizing effect, but are effective if associated with low light flux or darkness, as in *Chrysanthemum, Arabidopsis, Lactuca serriola*, and pea.[272-275]
3. High temperature devernalization has not been detected in a minority of species, e.g., *Geum, Lunaria annua*, and *Cardamine pratensis*.[31,104,276]

Devernalizing temperatures are usually in the 25 to 40°C range, but temperatures as low as 18 to 25°C may be effective.[21] Vernalizing and devernalizing temperatures are separated, in many plants by only a very narrow range of "neutral" temperatures, e.g., 13 to 15°C in Petkus rye, 17 to 18°C in sugarbeet, and about 20°C in *Hyoscyamus* and *Arabidopsis*. The extent of devernalization increases with the duration of the heat treatment (Table 2), at least up to a certain limit, e.g., few days at 30 to 35°C in winter cereals.[21,254]

Two major prerequisites must be met if heat devernalization is to be observed. *First*, the chilling treatment must be suboptimal. Indeed, optimally vernalized plants of several species, e.g., winter rye (Figure 3), cannot be devernalized.[254] In young *Arabidopsis* plants, however, devernalization is possible immediately after vernalization even when it is of optimal duration.[273]

Second, the heat treatment must follow immediately after the cold treatment. An

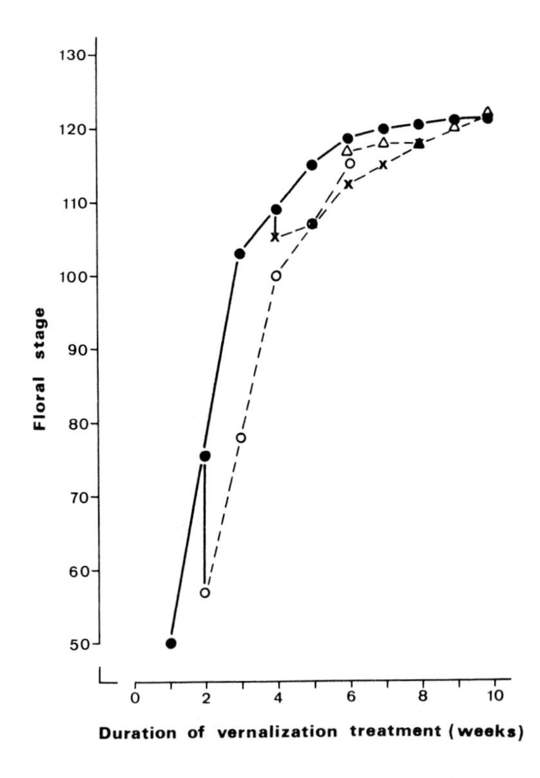

FIGURE 3. Course of revernalization after partial devernalization in *Secale cereale* var. Petkus (winter rye). All grains are first vernalized (●-●). After 2-, 4-, and 6-week vernalization samples of grains are exposed to 35°C for 3 days and then revernalized. The vertical lines give the immediate loss of vernalization and the broken lines the course of its restoration. The parallelism between the first and second vernalization processes suggests that revernalization is a repetition of vernalization. (From Purvis, O. N., *Encyclopedia of Plant Physiology,* Vol. 16, Ruhland, W., Ed., Springer-Verlag, Berlin, 1961, 76. With permission.)

intervening period at a neutral temperature apparently stabilizes the effect of the preceding chilling treatment and prevents any subsequent devernalization.[21,254,271,273] This may not hold true for all plants since in *Cheiranthus allionii* devernalization by heat becomes easier 1 or 2 months after chilling than immediately after chilling (Table 2) (see also Reference 275). In some plants the period at moderate temperature required for the so-called "stabilization" of thermoinduction might simply correspond to the time necessary for reaching the stage of initiation of flower primordia at the meristem, after which devernalization would of course be impossible.[240]

As a rule heat-devernalized seeds or plants can be revernalized by a further period of chilling, showing that both vernalization and devernalization are reversible processes (Figure 3).

Besides detrimental effects, heat may also promote flowering in some cold-requiring plants. This is exemplified by the remarkable case of *Scrofularia alata,* described by Chouard and Larrieu, a plant which flowers after either 6 weeks at 3°C or 3 weeks at 32°/27°C, but remains indefinitely at the vegetative rosette stage at 17°C.[277] In this case, heat may totally substitute for vernalization, a situation also found in *Festuca arundinacea.*[278] In this species, 45 days at 2°C or 45 days at 27°C are both below the

Table 2
DEVERNALIZING EFFECT OF 1 OR 2 WEEKS AT 35°C GIVEN AT INCREASED TIME INTERVALS AFTER SEED VERNALIZATION[a] IN *CHEIRANTHUS ALLIONII*

Time interval[b] between end of cold treatment and start of heat treatment (weeks)	% of flowering plants		
	No heat treatment	1 week at 35°C	2 weeks at 35°C
Control	56		
0		59	28
1		49	11
2		19	14
3		30	0
4		17	0
5		19	7
6		8	0
7		8	0

[a] Seeds are given a vernalization of 6 weeks at 5°C.
[b] Time intervals consist of LD given at 20°C.

From Barendse, G. W. M., *Meded. Landbouwhogesch. Wageningen,* 64(14), 1, 1964. With permission.

threshold for induction, but when these two treatments are given successively all the plants flower. Thus, the effects of the two subthreshold treatments may be summated even though the inducing factors used in these treatments are apparently opposite. A similar situation was described in the case of photoinduction in *Silene* (Volume I, Chapter 3, Table 2).

B. Light Conditions

Light conditions prevailing before, during, or after the chilling period may influence the effectiveness of the latter. These interactions are very intricate and diverse, and no attempt will be made here to cover all cases.

1. Before Vernalization

Before the cold treatment, good light conditions, i.e., high photon flux and/or LD, have usually a marked promotive influence in biennials and perennials, an effect attributable to perhaps a shortening of the juvenile phase in some species (see Volume I, Chapter 7, Section IV) and certainly to a better nutritional status of the plants as they enter the chilling conditions.

2. Short-Day Vernalization

In some cold-requiring plants, exposure to SD at temperatures above the vernalizing range has a clearly favorable influence. To summarize the SD-chilling interactions is difficult because they are complex. Species, however, tend to fall into one of two broad categories: first, SD may substitute partly or completely for chilling, a phenomenon called "SD vernalization"; second, SD and chilling are both required. A list of representative cases of the first group includes certain, but not all, varieties of winter cereals[21,250] and *Symphyandra hoffmani.*[279] Recall here that low temperatures may similarly substitute for the SD requirement in some SDP and SLDP (Volume I, Chapter 3). Species belonging to the second category are *Brassica oleracea* var. *gongyloides* (kohlrabi),[21] and some perennial grasses, such as *Poa pratensis* (Kentucky bluegrass),[280] *Bromus inermis* (Bromegrass),[281] and *Dactylis glomerata* (orchard grass).[282] In some of these plants, e.g., orchard grass, the two required factors can be clearly separated with the treatment by SD at moderate temperatures preceding the low temperature exposure.

More recent work of Blondon with *Dactylis* and of Heide with *Poa* indicates, however, that SD and low temperatures are independently capable of causing flower initiation in these perennial grasses.[244,283] Accordingly, *Poa* and *Dactylis* do not seem to behave very differently from the plants listed in the first category.

Whether SD and low temperatures activate the same or alternate pathways to induction is entirely unknown.

3. After Vernalization

A majority of cold-requiring plants also require LD after vernalization, this requirement being absolute in biennial *Hyoscyamus, Oenothera biennis,* and beet, and facultative in *Digitalis purpurea, Dianthus caryophyllus,* and *Teucrium scorodonia.*[21,112,284] Cultivars of *Chrysanthemum* behave as SDP and other species, e.g., *Scrofularia alata, Geum urbanum, Lunaria annua,* and *Dianthus barbatus,* as DNP after completion of vernalization.[104,112,271] In *Cheiranthus allionii,* the requirement for LD after thermoinduction is absolute when this species is vernalized at the seed stage, decreases markedly when chilling is applied at later stages of growth, and disappears totally in old vernalized plants.[240]

The photoperiodic requirements of plants are markedly changed after thermoinduction. The data of Figure 1 suggest that as vernalization proceeds, fewer LD are required for induction. In some species, thermoinduction may even suppress the requirement for LD. The case of the biennial *Melilotus officinalis* (sweet clover) is an excellent illustration of this situation. After vernalization at 7°C for 4 weeks, this species flowers in days of 13 hr or more, whereas unvernalized plants flower only in photoperiods of 20 hr or more (Table 3). In temperate zones, 20-hr daylengths being unnatural, the plant is classified as cold-requiring, but in arctic regions where summer days of 20 hr are common, the same plant is a LDP with a long, critical daylength.[285] The behavior of *Melilotus* is probably not exceptional, and several so-called LDP may become DNP when vernalized (see Volume I, Chapter 3, Section VIII. A; also Reference 286).

On the other hand, in several biennials, e.g., sugarbeet,[284] *Oenothera,*[241] and *Cheiranthus,*[240] the thermoinduced state may disappear more or less rapidly if the plants are grown in SD at moderate temperatures after the chilling treatment (Figure 4). The complexity of interpreting environmental signals is emphasized: SD that substitute for the chilling treatment in some species (see above) produce a devernalizing effect in others.

VI. INDUCED VERNALIZABILITY

Under certain environmental conditions one can demonstrate a favorable response to low temperature where none existed before. This becomes evident in *spring* rye after germination under anaerobiosis,[287] in noncold-requiring strains of *Arabidopsis* grown under low light flux or SD conditions,[288] and in pea after amputation of the cotyledons.[289] Such complex phenomena are difficult to interpret without additional information on the nutritional status of the plants. It is clear that low temperatures in reducing growth and respiratory losses and promoting the breakdown of starch and other reserves may indirectly improve the assimilate supply to shoot apices and, in that

Table 3
INTERACTION BETWEEN VERNALIZATION AND PHOTOPERIOD IN *MELILOTUS OFFICINALIS*

Vernalization[a] temperature (°C)	Days to flowering Photoperiod[b] (hr)		
	13	15	17
7	81	78	63
10	∞[c]	∞	75
15	∞	∞	88
20	∞	∞	∞

[a] Vernalization for 4 weeks.
[b] Photoperiods given at 20°C.
[c] Plants still vegetative at the end of the experiment.

From Kasperbauer, M. J., Gardner, F. P., and Loomis, W. E., *Plant Physiol.*, 37, 165, 1962. With permission.

fashion, promote their progress towards the reproductive condition (see Volume II, Chapter 8).

VII. STABILITY OF THERMOINDUCTION

Fractional thermoinduction, i.e., summation of two subthreshold vernalizing treatments, is not as common as fractional photoinduction because of the variety of devernalizing environmental parameters, e.g., high temperatures, low irradiances, and SD. A successful fractionation of thermoinduction was reported in the biennial *Lunaria annua*.[290] This plant does not respond to seed vernalization, and its "vernalizability" increases with age once a juvenile phase of about 6 weeks has been completed. Two chilling treatments, one applied to imbibed seeds and the other to plants of various ages, can be summated even with several weeks interpolated (Table 4). The effect of the first treatment, in itself insufficient to produce a floral response, thus persists in this case during quite a prolonged period of nonvernalizing conditions.

The thermoinduced state, if completely established, is also highly stable in some other species. Thus, in biennial *Hyoscyamus*, with an absolute requirement for cold and LD, vernalized individuals maintained continuously in SD retain for at least 200 days the capacity to flower in response to a transfer in LD, that is the thermoinduced state persists for at least 7 months.[21,245] After 300 days, however, there is a decrease in thermoinduction. Fully thermoinduced plants of *Lolium perenne* and *Hordeum bulbosum* held in SD for weeks remain vernalized and able to initiate flowers in subsequent LD.[291,292]

Also, vernalized seeds of *Sinapis* and *Cheiranthus* (but not those of winter rye) can be redried and stored at room temperature for extended periods, up to 6 years in the case of *Sinapis*, without any resultant devernalization.[240,293,294]

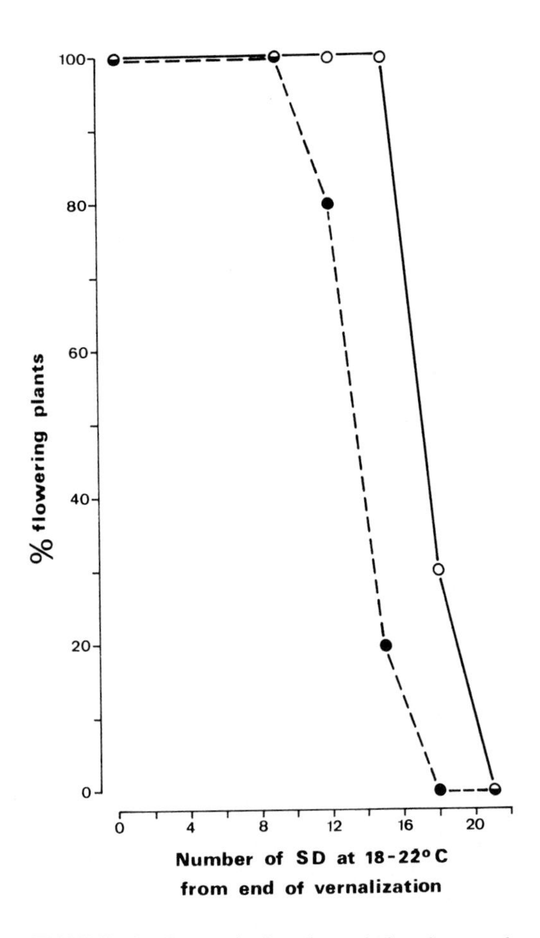

FIGURE 4. Loss of the thermoinduced state by growth in SD after an effective vernalization in the biennial *Oenothera biennis*. ●---●, vernalized during 13 weeks; o — o , vernalized during 16 weeks. After the SD treatment, the plants are exposed to LD which are absolutely required for bolting and flowering in this species. (Adapted from Picard, C., *Ann. Sci. Nat., Bot. Biol. Veg.*, 12e sér., 6, 197, 1965.)

Transmission, and thus persistence, of the vernalized condition throughout growth and development of the plants occurs apparently during the process of division of the meristematic cells originally exposed to cold. Indeed, by removing the main shoot of fully vernalized Petkus rye, so that the tillers are stimulated to grow, and by removing these to stimulate secondary tillers and so on, Purvis has observed that even 4th order tillers flower normally.[249,259] The plants remain vernalized although the meristems of the 4th order tillers were nonexistent at the time of chilling treatment. Similar experiments by Schwabe in *Chrysanthemum* and by Barendse in *Cheiranthus,* involving one or more decapitations, up to seven in the case of *Chrysanthemum,* have yielded identical results.[240,259] One concludes, therefore, that thermoinduction of the growing shoot apex can be transmitted without any apparent loss to lateral meristem which are formed after vernalization of the original meristems.

The effect of vernalization appears quite immobile in many species since the flowering response is localized to the portion of the plant exposed to low temperature,[112,243,260,282] and since there was no evidence indicating translocation of a vernalization stimulus in grafting experiments.[31,240,241,259] In *Cheiranthus,* removal of the

Table 4
INFLUENCE OF THE AGE OF THE PLANT AND OF A PRECEDING SEED VERNALIZATION ON THE RESULT OF PLANT VERNALIZATION IN THE BIENNIAL *LUNARIA ANNUA*

Age in weeks at the start of plant vernalization[a]	% of flowering plants	
	Without seed vernalization	With seed vernalization[b]
4	0	0
6	0	5
8	65	95
10	88	100
12	100	100

[a] Plant vernalization: 11 weeks at 5°C.
[b] Seed vernalization: 6 weeks at 2°C. This treatment alone never results in flower initiation.

Adapted from Wellensiek, S. J., *Proc. K. Ned. Akad. Wet.*, 61, 561, 1958.

plant part that is derived from the originally vernalized apex totally prevents flowering of the laterals that develop afterwards.[240] Thus, *only the axillary meristems formed from a vernalized meristem attain a vernalized condition.*

At first sight stability of thermoinduction is unexpected because of the continuing "embryogeny" characteristics of angiosperm meristems, which apparently renew themselves continuously. The nature of the meta-stable change affecting the cells of vernalized shoot apices is completely unknown. This change is reversible; however, since (1) continuous removal of flower buds on floriferous stems of *Geum* for 1 year leads to a progressive loss of the vernalized state;[31] and (2) winter annual or biennial plants arising from seeds produced by vernalized parents require fresh vernalization. The thermoinduced state is not carried from one generation to the next, and the moment when this state is lost is often believed to coincide with the time when meiosis occurs. In pea, however, Reid found that vernalization of the parents causes a small, but significant promotion of flowering of the progeny, i.e., it could be transmitted through a meiotic division.[295] This effect disappears in the next generation.

Thus, the stability of the thermoinduced state may differ greatly depending on the species considered. A similar situation was reported in photoperiodic plants (Volume I, Chapter 3). The differences between different species remain totally unexplained.

CONCLUSIONS

Many plants exhibit a requirement for low-temperature treatment (vernalization) before they are induced to initiate flower primordia. As a rule, plants having a facultative requirement (winter annuals) can be vernalized as imbibed seeds, whereas those with an obligate requirement (biennials, perennials) cannot and must reach a certain size for attainment of responsiveness to cold. Many experiments show that the shoot apex is the site of perception of the vernalization stimulus and suggest that dividing (or cycling) cells must be present if thermoinduction is to occur.

There is a great and subtle variety of interactions between vernalization and other environmental factors. Exposure to high temperatures, low light flux, or SD, immediately following low-temperature treatment can prevent thermoinduction (devernalization) in some species and under some circumstances. Devernalized plants can be revernalized showing that both vernalization and devernalization are reversible processes.

SD, that produce a devernalizing effect in some species, may on the contrary substitute for the chilling treatment in others.

In many plants, the photoperiodic requirements are reduced or even altogether suppressed after thermoinduction.

The thermoinduced state, if completely established, may be highly stable, although not irreversible, in some plants. Transmission of the vernalized condition throughout growth and development of these plants occurs during the process of division of the cells originally exposed to cold. Thus, only the axillary meristems derived from a vernalized meristem attain a vernalized condition. In view of this, stability of thermoinduction could be different from that of photoinduction (see Volume I, Chapter 3). In other cold-requiring species, because of the variety of devernalizing factors, the thermoinduced state is far less persistent.

It is emphasized that one must clearly separate thermoinduction and photoinduction, not merely from the obvious point of view of the parameters involved, but more importantly from considerations of the receptor tissues and the kinds of internal changes that must occur as a result of each.

Chapter 6

CLASSICAL THEORIES OF INDUCTION

TABLE OF CONTENTS

I. INTRODUCTION

When one considers the enormous diversity of plant responses to daylength, temperatures, and other environmental parameters (Volume I, Chapters 2, 3, and 5), it is not easy to see how to reconcile all the data into a single theory. This pessimistic view is not shared, however, by many flowering physiologists and a number of "universal" theories have been enunciated during the last half-century. As pointed out by Lang, two basically different possibilities can be visualized in photoperiodic and cold-requiring plants: (1) inductive conditions promote flower initiation; and (2) noninductive conditions inhibit it.[296]

Alternative *1* implies that plants are incapable of flower initiation unless they are induced, whereas *2* implies that plants are potentially capable of flower initiation, but are secondarily suppressed by noninductive conditions.

Alternative *1* gave rise to the concept of florigen which is the dominant theory in the field.

II. THE FLORIGEN THEORY

By observing that leaf cuttings prepared from flowering *Begonia* plants produce adventitious shoots which promptly flower, whereas leaf cuttings taken from vegetative plants regenerate vegetative shoots, Julius Sachs in the 19th century was probably the first to support the notion of "flower-forming substances" present in flowering plants. He also regarded the generation of other organs, e.g., roots, leaves and shoots, as under the control of specific organ-forming substances. Nearly 60 years later, when it was found that photoperiod is perceived by leaves and that photoperiodic plants flower irrespective of the daylength conditions to which the apical meristem is exposed, it became obvious that some information is transmitted from the leaves, which causes a floral response in the meristem.[68] In 1937, Chailakhyan, not believing in electrical signals in plants, proposed that the leaf-generated signal is a substance of hormonal nature, viz., "florigen".[297] This concept, because it received some support from grafting and other experiments, was soon widely accepted and has been "flourishing" ever since.

A. Experimental Evidence Concerning Florigen

1. Grafting Experiments

The basic evidence for the existence of florigen comes from numerous grafting experiments in which a "receptor" plant kept in noninductive conditions is induced to flower by union with a previously induced "donor" plant (Tables 1 and 2). Transmission of the floral stimulus through a graft union has been observed in experiments using donor and receptor plants of the same species or using two different species or genera as donor and receptor, even when the two partners belong to different photoperiodic groups (Figure 1). In some instances, e.g., in tobaccos, soybean, *Perilla,* and *Kalanchoe,* the donor can be reduced to a single induced leaf. Appropriate controls, i.e., noninduced control grafts, have not often been made in these experiments because they are technically difficult to perform. However, when made, they have generally indicated that transmission is obtained only if the donor is induced. In species, like *Silene* and *Perilla,* in which alternate environmental pathways to flower initiation have been found (cold or warm SD in the LDP *Silene;* cold LD in the SDP *Perilla*), all kinds of induced plants can transmit flowering to a vegetative graft partner.[97,98] Deronne and Blondon have also shown that the summation of two different subthreshold inductive treatments in *Perilla* results in the production of a transmissible hormone indistinguishable from that produced after a regular induction by SD. All these results

Table 1
SOME SUCCESSFUL CASES OF TRANSMISSION OF THE FLORAL STIMULUS IN GRAFTING EXPERIMENTS INVOLVING MEMBERS OF THE SOLANACEAE[21,71,298]

Donor[a]	Response type	Receptor[b]	Response type
Nicotiana sylvestris	LDP	Maryland Mammoth tobacco	SDP
Trapezond tobacco	DNP	*Nicotiana sylvestris*	LDP
Trapezond tobacco	DNP	Maryland Mammoth tobacco	SDP
Maryland Mammoth tobacco	SDP	*Nicotiana sylvestris*	LDP
Maryland Mammoth tobacco	SDP	*Hyoscyamus niger* (annual)	LDP
Hyoscyamus niger (annual)	LDP	Maryland Mammoth tobacco	SDP

[a] Donor is always induced to flower, often showing flower buds.
[b] Receptor is always kept vegetative by growth in noninductive conditions.

Table 2
SUCCESSFUL (+) AND UNSUCCESSFUL (−) CASES OF TRANSMISSION OF THE FLORAL STIMULUS IN GRAFTING EXPERIMENTS INVOLVING MEMBERS OF THE CRASSULACEAE[71,91,299]

	Donor (induced plant)			
Receptor (vegetative plant)	*Kalanchoe* (SDP)	*Sedum* (LDP)	*Bryophyllum* (LSDP)	*Echeveria* (SLDP)
Kalanchoe blossfeldiana (SDP)	+	+	+	+
Sedum spectabile (LDP)	+	−	Not tried	Not tried
Bryophyllum daigremontianum (LSDP)	±	−	+	Not tried

have generated great enthusiasm among flowering physiologists because they clearly support the concept of a transmissible floral hormone and because they tend to indicate that this hormone is identical, at least physiologically (often assumed to mean chemically), in all photoperiodic species, including DNP.[21]

Transmission has been obtained in many cases whether the donor is the stock or the scion showing that the movement of the floral stimulus is basically possible in both the upwards and downwards directions. This movement is in general greatly facilitated by removal of leaves from the receptor and buds from the donor.[21] Removal of donor leaves usually prevents transmission presumably because the source of stimulus is suppressed.

Zeevaart devised a very simple and elegant system to study the transmission of floral stimulus in the SDP *Perilla*.[71] Induced leaves (donor leaves), with blade area reduced to 30 cm^2, are grafted by inserting the petiole into a young internode at the top of a decapitated vegetative plant. The stock plant is totally defoliated and disbudded, except for one leaf pair at the base of the plant and the two axillary buds of the uppermost node which are used as receptor buds. In this system, the minimum duration of contact between an optimally induced leaf and the stock plant in order to get a floral response is the same as that necessary to detect the first translocation of [^{14}C] sucrose through the graft union. The transmission of floral stimulus and that of assimilates seem to occur simultaneously and thus require the establishment of phloem continuity between the two partners of the graft. Very similar results are obtained in the LDP,

FIGURE 1. Transmission of the floral stimulus through a graft union between a stock of the LSDP *Bryophyllum daigremontianum* and a scion of the SDP *Kalanchoe blossfeldiana.* The composite plant is kept in LD after grafting. Left, both partners were noninduced prior to grafting. Right, the *Bryophyllum* stock was induced prior to grafting (inflorescence was removed at the time of grafting) and causes abundant flowering of the *Kalanchoe* scion in LD. (From Van de Pol, P.A., *Meded. Landbouwhogesch. Wageningen,* 72(9), 1, 1972. With permission.)

Silene armeria, when the donor plant is used as stock and the receptor as scion. Four days are then necessary for the first upwards transport of both the stimulus and [^{14}C] assimilates.[53] However, in the reverse combination, when the donor is used as a scion, it takes about 7 days for the first downwards transport of [^{14}C] assimilates and at least 3 weeks for downwards transmission of the floral stimulus.[97,300] The reason for this difference is unknown, but such a result shows that the grafting technique may sometimes make interpretation of the results less than straightforward.

We have seen (Volume I, Chapter 3, Section VII) that plants of *Xanthium, Bryophyllum,* and *Silene* possess the unique property of indirect induction. Apparently, these plants share the same floral stimulus with species devoid of this property, as shown for example in graft combinations between *Bryophyllum* and *Kalanchoe* (Table 2; Figure 1). However, the attribute of indirect induction cannot be transmitted to other species by grafting.[91]

The results of grafting experiments with biennial *Hyoscyamus* support the idea that a transmissible stimulus that Melchers called "vernalin", is formed as a result of low-temperature and LD treatment.[301] Similar cases of transmission of a floral stimulus through a graft union were reported in few other thermophotoinduced biennials, e.g.,

Table 3
SOME SUCCESSFUL CASES OF TRANSMISSION OF THE FLORAL STIMULUS IN GRAFT COMBINATIONS BETWEEN PHOTOPERIODIC AND COLD-REQUIRING SPECIES[21,261]

Donor (induced plant)	Response type	Receptor[a]	Response type
Hyoscyamus niger (annual)	LDP	*Hyoscyamus niger* (biennial)	CRP
Maryland Mammoth tobacco	SDP	*Hyoscyamus niger* (biennial)	CRP
Beta vulgaris (annual)	LDP	*Beta vulgaris* (biennial)	CRP
Brassica crenata (annual)	LDP	*Brassica napus* (winter annual)	CRP
Anethum graveolens	LDP	*Daucus carota* (biennial)	CRP
Sinapis alba	LDP	*Brassica oleracea* (biennial)	CRP
Shuokan *Chrysanthemum*	CRP	Honeysweet *Chrysanthemum*	SDP
Hyoscyamus niger (biennial)	CRP	Maryland Mammoth tobacco	SDP

[a] Receptor is kept vegetative by growth in noninductive conditions. In the case of cold-requiring plants (CRP), the nonthermoinduced receptor is kept in LD.

beet, carrot, and cabbage.[21] This is not necessarily evidence, however, for transport of a stimulus formed as a result of cold treatment *alone*.[21,302] Schwabe,[259] working with a cold-requiring cultivar of *Chrysanthemum*, found no evidence indicating translocation of the vernalization stimulus and similar negative results were obtained in perennial grasses,[243,282] *Cheiranthus*,[240] olive,[260] *Geum*,[31] and *Oenothera*.[241] As noted in Volume I, Chapter 5, Section VII, the immediate product of vernalization seems quite immobile in most plants.

The question of the relationship between vernalin and the floral stimulus of photoperiodic species has been investigated using grafting techniques (Table 3). Induced LDP or in one case an induced SDP are effective donors for nonthermoinduced cold-requiring plants. And, on the other hand, thermoinduced biennial *Hyoscyamus* and Shuokan *Chrysanthemum* are found to be effective donors for vegetative receptors of the SDP, Maryland Mammoth tobacco, and Honeysweet *Chrysanthemum*, respectively.

All these results seem to indicate that the floral hormone is the same in all higher plants. However, there are a number of complications and anomalies in some grafting experiments that seem to preclude such a generalization.

First, there is an ever increasing list of unsuccessful cases of transmission, even in expert hands, from an induced donor to a vegetative partner. The DNP Delcrest tobacco is an effective donor for the LDP *Nicotiana sylvestris*, but not for the SDP Maryland Mammoth tobacco.[71] Similarly, the LDP *Blitum capitatum* is an excellent donor for receptors of its own species, but not for those of the LDP *Blitum virgatum*.[303] There is no transmission between *Cestrum diurnum* and *C. nocturnum*, but there is from *C. diurnum* to *C. diurnum* across a bridge of *C. nocturnum*.[304] *Silene armeria* can only transmit the floral stimulus to receptors of *S. armeria*, but not to those of *S. gallica, S. otites, S. nutans,* and *Melandrium album*.[91] After an extensive search for new successful cases of transmission among members of the Caryophyllaceae, Van de Pol concludes that interspecific and intergeneric transmission of the floral stimulus is the exception rather than the rule. Very curious, too, is the failure of induced plants to transmit the floral stimulus to a vegetative receptor of their own species, as found in various *Sedum, Silene cucubalis, S. nutans, S. otites, Melandrium album,* and *Blitum virgatum*.[71,91,303] An explanation for a part of these failures is that the induced donor reverts to vegetative growth after return to noninductive conditions at the time of grafting, but this explanation certainly does not apply to all these cases.

Table 4
CASES OF FLOWER INITIATION IN GRAFTS BETWEEN 2 VEGETATIVE PARTNERS[21,71,296]

Donor[a]	Response type	Receptor[a]	Response type
Maryland Mammoth tobacco	SDP	*Hyoscyamus niger* (biennial)	CRP
Hyoscyamus niger (annual)	LDP	*Hyoscyamus niger* (biennial)	CRP
Maryland Mammoth tobacco	SDP	*Nicotiana sylvestris*	LDP

[a] In these cases, the donor is maintained strictly under noninductive conditions and is thus vegetative. The term "donor" is justified because grafting of this plant to a partner also kept in noninductive conditions produces flowering of the latter, which is then called the "receptor".

Second, several examples of nonreciprocal transmission are known. Shuokan *Chrysanthemum, Bryophyllum daigremontianum,* and *Chenopodium rubrum* are good donors for Honeysweet *Chrysanthemum, Kalanchoe blossfeldiana,* and *Blitum capitatum,* respectively. However, in reciprocal graft combinations, *Kalanchoe* and *Blitum* have been proved to be very poor donors for *Bryophyllum* and *Chenopodium,* respectively, while Honeysweet *Chrysanthemum* is a totally ineffective donor.[91,261,303] An explanation often offered for this kind of anomaly is that the capacity of florigen production is higher in some species than in others or, alternatively, a lower level of the hormone is needed for flower initiation in some species than in others.[91,92,303] Other explanations, however, are equally plausible and will be discussed below.

Third, a *noninduced* SDP or a *noninduced* LDP are effective donors for nonthermoinduced biennial *Hyoscyamus* (Table 4) suggesting that vernalin, the transmissible hormone postulated in some cold-requiring plants is different of florigen, the floral stimulus of induced photoperiodic plants. Melchers assumes that vernalin is a "physiological" precursor of florigen,[301] but since a *noninduced* SDP is also an effective donor for a noninduced LDP (Table 4), the general impression gained from these observations is that there are different floral hormones in different species.

Taken together, all these negative and anomalous results of grafting experiments leave *real doubts about the existence of a floral hormone common to all plants.* On the other hand, because grafts can only be made between compatible species, the grafting technique is totally unable to demonstrate the universality of this hormone.

By leaving the receptor in contact with an incompatible donor for only a few weeks, after which the receptor is regrafted onto a vegetative stock of its own species, Wellensiek and Van de Pol have tried to overcome the problem of incompatibility.[91,305] Using this technique, they have obtained transmission of the floral hormone from an induced *Xanthium* (Compositae, SDP) or an induced *Perilla* (Labiatae, SDP) to vegetative *Silene* receptors (Caryophyllaceae, LDP). As no vascular connections are formed in these incompatible grafts, transmission of the floral stimulus must have occurred from cell to cell. A *Silene* donor is, however, unable to cause flower formation in a *Xanthium* receptor. Likewise, there is no transmission in several other incompatible combinations performed with the same technique,[91] so that the significance of the two positive results so far obtained in such grafts remains problematical.

Finally, we wish to emphasize that the grafting technique is obviously entirely inappropriate to yield decisive information about the nature — simple or complex, specific or unspecific — of the transmissible stimulus.

2. Parasitic Plants

The connection between a parasitic plant, like *Cuscuta* (dodder), and its host plant has many analogies with a graft union between two different species. The parasite manages to make connections with the phloem elements of the host and becomes then totally dependent on the latter for food supply. As florigen is said to move in the phloem with assimilates, flower initiation in the dodder should be dependent on the induction of the host. Contrary to expectation, flowering dodder have been found on flowering hosts as well as on hosts that are in a strictly vegetative condition.[79,306] On the other hand, there is no transmission of the floral stimulus from a flowering dodder to a defoliated vegetative host, but the movement of any material from parasite to host might be totally prevented even when the latter is starved. The results are at best inconclusive concerning the existence of florigen. The two partners appear quite independent one from the other concerning their own progress towards the reproductive condition. This conclusion will have to be modified in a later section, however, when inhibitors of flower initiation will be discussed.

3. *Studies on the Transport of the Floral Stimulus in a Single Plant*

By definition a hormone is a translocated substance, and hence, many studies were initiated in order to determine the time, velocity, and pattern of transport of the postulated floral hormone. Needless to say that such an approach, because the physical and chemical nature of this hormone is unknown, can only be indirect and requires ingenious, perhaps also ingenuous, experimenters.

a. Timing and Velocity of Transport

The time of movement of florigen out of the induced leaves in photoperiodic species reacting to a single inductive cycle can be determined by defoliating different groups of plants at various intervals following the start of the inductive cycle. This kind of experiment, first performed on *Xanthium*,[73] was later extended to other species (Figure 2). In all cases, there is a critical time before which leaf removal totally suppresses flower initiation and after which the same treatment no longer prevents it. This is taken as indicating the movement of a sufficient quantity of florigen out of the induced leaves. The time at which this delivery starts is remarkably early in most plants, i.e., 16 to 24 hr after the start of the inductive cycle. These data suggest that the floral hormone moves out of the leaves soon after it is formed and is translocated only over a short period following induction. They give no clue, however, about the nature of this hormone nor about its immediate destination either the receptor meristem or other plant parts, and may even be misleading concerning the time of its export out of the leaves. Suppose this hormone is complex, the method of successive defoliations tells only of the movement of the slowest, or last, component. The fastest, or first, component(s) may have moved undetected far earlier.

The technique of sequential defoliation was further refined by removing the source of florigen, not only at various times, but also at several points along the translocation pathway. In this way the times at which enough hormone to cause a particular level of floral response pass several points of known distance apart can be determined, and an estimate of the velocity of its movement can be tentatively calculated. With this method, Evans and Wardlaw obtained a velocity of 1 to 2 cm/hr in the LDP *Lolium*,[309] whereas Takeba and Takomoto and King et al. arrived at a much higher value of 30 cm/hr or above in the SDP *Pharbitis*.[310,311] At present it is not possible to decide whether this difference is because the movement was followed in the leaf blade in *Lolium* and in the stem in *Pharbitis*, or because the hormone or a component of it is different in different species. If we had specific radioactive markers for the floral stimulus(i), as we do for other compounds, we could then really estimate the velocity of "frontal" movement for this stimulus. In the absence of such markers, the interpretation given for the above data is questionable and all estimates dubious.

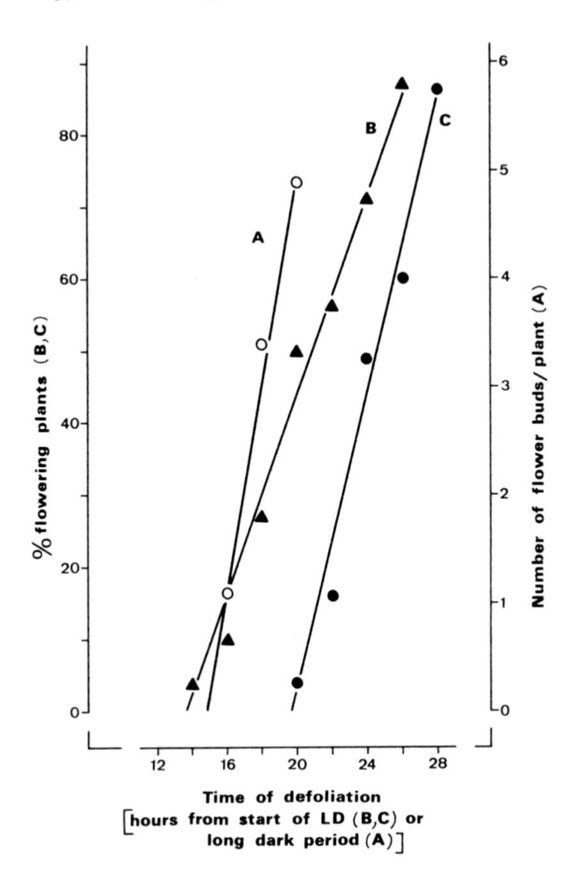

FIGURE 2. Export of the floral stimulus out of the induced leaves in plants reacting to a single inductive cycle, as studied by the sequential removal of the induced leaves at various time intervals after the start of the inductive cycle. A, *Pharbitis nil* (SDP); B, *Sinapis alba* (LDP); and C, *Lolium temulentum* (LDP). Plants of *Pharbitis* are induced by one 16-hr long night and plants of *Sinapis* and *Lolium* by one LD of 22 and 32 hr, respectively.[307-309]

b. Pattern of Transport

Early work with photoperiodic species indicates that the floral hormone moves only through living tissues.[21] Within the leaf blade movement probably occurs from cell to cell through the mesophyll until loaded into the phloem. Further transport in the petiole and stem is apparently in the phloem since it can be prevented by treatments that interrupt this tissue mechanically or physiologically, e.g., girdling, localized cold or heat treatment, and narcotic application.[1,21,71,312] It was also found that florigen is incapable of traversing a water- or an oil-gap between an induced and a noninduced plant part.[313,314] The last step of its travel towards the target meristem is probably from cell to cell beyond the ends of the protophloem strands, over a distance which can be estimated at about 0.5 mm in many species.

Assimilates are known to move in phloem by mass flow from sources to sinks. Studies with $^{14}CO_2$ show that the pattern of assimilate distribution from any one source, for example a photosynthesizing leaf, is dependent not only on the number and localization of sinks or regions of utilization, but also on the supply of assimilates from other sources. Generally the ability of different buds to function as sinks differ greatly according to their growth rate. In many herbaceous plants the shoot apical bud is growing most actively and is a greater attractive force for assimilates than the less active axillary buds. In a growing bud, young expanding leaves are competing and very active sinks, probably far more effective than the meristem itself in terms of absolute demand for assimilates. Removal of particular sinks, like some buds or young leaves, results in the redirection of assimilates to other sinks which then receive a better supply. Removal of selected mature leaves (sources) may similarly alter the distributional pattern of assimilates.

If florigen is moving passively with assimilates its transport must be influenced by all factors that modify the translocation of these assimilates. This is indeed what is observed in a number of experiments. In several species, using two-branched plants in which one branch is exposed to inductive cycles (donor branch) and the other to noninductive cycles (receptor branch), it is easy to demonstrate that the floral hormone can move basipetally in the former branch and acropetally in the latter, i.e., its movement is nonpolar as is the flow of assimilates.[21] As in grafting experiments, this movement of florigen from donor to receptor branch is depressed if the receptor possesses leaves.[315] The inhibitory effect of noninduced leaves in many species, e.g., *Perilla, Sinapis, Kalanchoe,* and *Bryophyllum,* is essentially restricted to leaves that are (1) mature and 2) located between the induced leaves and the receptor meristem, i.e., leaves that presumably provide an alternate source of assimilates and thus prevent the floral hormone from reaching the receptor meristem. Removal of these leaves greatly improves flowering.[21] The same effect is obtained by removal of the growing reproductive buds or young leaves on the donor, i.e., buds or leaves which most probably act as competing sinks for the hormone.[315]

In plants with an opposite decussate phyllotaxis, the floral hormone preferentially moves through the stem section which is adjacent to the induced leaf. The only inhibitory leaves in these plants are essentially those inserted in the same orthostichy as the induced ones. Also, one-sided flowering in *Perilla* and inflorescences with unilateral flowering in *Kalanchoe* and *Bryophyllum* are obtained with one-sided induction, i.e., when only one leaf of a pair is induced.[69,316,317] Apparently, lateral movement of florigen within the stem of these plants is almost totally prevented, probably due to lack of efficient lateral connections in their vascular system. Indeed, studies on the distribution of dyes or labeled assimilates in the same plants have shown that the flow of substances through the stem is mostly restricted to the vascular tissue that is in most intimate contact with the leaf from which these substances are derived.[318]

All the above evidence indirectly indicates that the distribution of florigen in the plant is dependent on the pattern of distribution of the assimilates originating from the induced leaves. More recent work in the SDP *Perilla* by Chailakhyan and Butenko and by King and Zeevaart was aimed at testing this assertion.[318,319] The distribution of pulse-labeled assimilates, as determined by autoradiography or by radioactivity counting, and that of florigen, as determined by the spatial localization of reproductive buds, were compared in the same individuals. It was found that there is a relatively good correlation between the translocation into receptor meristems of [^{14}C] assimilates from the induced leaf and the movement of the floral hormone (Table 5). Lack of flowering in a particular meristem is invariably associated with the absence of transport of assimilates from the induced donor leaf into this meristem. Such a vegetative meristem receives assimilates predominantly from noninduced leaves. Flower initiation occurs in the meristem(s) that is essentially supplied by the induced leaf. Similar results were also obtained in two other SDP, *Pharbitis* and *Xanthium.*[311,312,320]

Translocation of photosynthates and transmission of the floral stimulus are not always correlated. In the LDP *Lolium* and *Anagallis* (GO strain) leaves, that are too young to export assimilates, are very effective in causing flower initiation, suggesting that the floral stimulus moves independently of assimilates in these species.[56,309]

Table 5
FLORAL RESPONSE AND [^{14}C] ASSIMILATES IMPORT IN AXILLARY SHOOTS (♀) OF THE SDP *PERILLA CRISPA* AS AFFECTED BY THE PRESENCE OF A PAIR OF LD LEAVES AT NODE 2

Shoot at node	Floral response and imported radioactivity	Left ♀	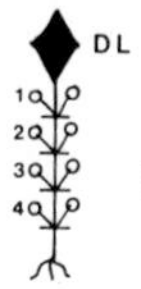 Right ♀	Left ♀	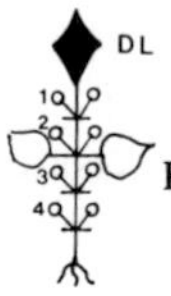 Right ♀
1	Flowering[a]	100	100	100	100
	Days[b]	13	13	12	15
	Radioactivity[c]	34,940	19,810	5,540	51,290
2	Flowering	100	100	66	71
	Days	18	19	45	37
	Radioactivity	29,070	28,010	0	30
3	Flowering	100	100	100	100
	Days	20	19	21	27
	Radioactivity	14,730	610	210	4,960
4	Flowering	100	100	66	50
	Days	24	21	49	46
	Radioactivity	810	120	0	0

[a] Flowering expressed as the percent of flowering plants.
[b] Days from grafting of the donor leaf (DL) to appearance of flower buds. The experimental set-up is similar to that described in Section II. A. 1.
[c] Imported radioactivity (in cpm) 2.5 hr after a 30-min labeling of the donor leaf with $^{14}CO_2$. Radioactivity is measured in 80% ethanol extractable material.

From King, R. W. and Zeevaart, J. A. D., *Plant Physiol.*, 51, 727, 1973. With permission.

4. Fractional Induction Experiments

The possibility of fractional induction has been taken by some researchers as supporting evidence for the specificity and uniqueness of the flower-promoting factor.[21] They believe that a specific substance, like the presumptive florigen, should have a better chance to persist through extended periods of noninductive conditions than a common metabolite or a particular ratio of two or several substances. The results of experiments on fractional induction are, however, so complex and diverse that it is hardly possible to reach any general conclusion (see Volume I, Chapter 3, Section VI), and indeed some of these experiments have been interpreted in terms of production of a floral inhibitor by leaves in noninductive conditions (see Section III. A. 3. below).

B. Identity of Florigen

1. Isolation

The most decisive test for the florigen theory would be the isolation of this hormonal principle. The basic requirements of a successful isolation could be defined as follows: (1) the active material should be present in flowering individuals and absent in vegetative individuals of the same species; (2) it should be common to flowering individuals of various species belonging to various photoperiodic groups; and (3) application of this material should cause flower initiation in vegetative individuals of different species belonging to various photoperiodic response types. After more than 40 years of extensive and careful work along these lines, we do not have the merest idea of its chemical structure. In *Xanthium* alone, thousands of attempts to isolate flower-inducing activity from extracts were performed in Bonner's laboratory, but these did not meet with any success.[95] From time to time, there have been reports of positive effects, but they seem to be for the most part irreproducible or obtained in growing conditions that permitted a marginal or low flowering in the controls.[93,321-323]

Some workers have had more consistent results. A description of their experiments illustrates the numerous difficulties that are encountered in this kind of study.

Extracts prepared from flowering plants in various laboratories were found to produce bolting and flowering in some vegetative rosette LDP, e.g., *Samolus parviflorus* and *Rudbeckia bicolor*.[21,261] Since (1) exogenous GA_3 has exactly the same effects (see Volume II, Chapter 6) and (2) gibberellin-like materials are present in these extracts, their activity can be ascribed to the presence of GAs. These compounds are not considered as floral hormones by most proponents of the florigen theory, however, because the GAs are not universal promoters of flower initiation (Volume II, Chapter 6).

Lincoln and co-workers prepared a crude methanol extract from lyophilized flowering branch tips of *Xanthium* and found that this extract causes some flowering in vegetative *Xanthium* receptors while an extract from vegetative plants is inactive.[324] Not all meristems of treated plants responded and those responding failed to develop beyond an early floral stage (about stage 4 of the scale depicted in Volume I, Chapter 1, Figure 2A). Further efforts to purify the active principle have been so far unsuccessful and fractionation of the extract beyond a certain limit leads inexorably to loss of activity.[325] The active material is apparently highly water soluble with the partitioning properties of a carboxylic acid, and it has been referred to as "florigenic acid".[326] The effectiveness of the crude extract was increased by Carr by addition of a small amount of GA_3 which by itself cannot produce a floral response.[327] Interestingly enough, the plants repeatedly treated by a combination of extract and GA_3 only produce phyllodic male inflorescences and no female inflorescence at all, suggesting that complete flowering in *Xanthium* may depend on several substances. If true, this would explain why activity of extracts is lost on purification and fractionation.

A crude acetone extract from flowering *Xanthium* buds is without florigenic activity when tested on vegetative *Xanthium* receptors, but is able to cause flower initiation in the SDP *Lemna paucicostata* 6746 grown in noninductive conditions.[328] When supplemented with GA_3 the acetone extract is effective in *Xanthium* and totally ineffective in *Lemna*! This work has not progressed very much in the last years mainly because positive results could not be obtained consistently in the hands of several workers.[92]

Another very elegant approach to this problem has been taken by Cleland. Knowing that the floral hormone is transported in the phloem and that aphids feed on phloem sap and produce honeydew that is qualitatively quite similar to phloem sap, this author undertook to comparatively analyze the honeydew collected from aphids feeding either on flowering *Xanthium* plants or on vegetative plants. He found that a fraction which is present in both flowering and vegetative honeydew causes flowering in the LDP *Lemna gibba* G3 under noninductive conditions.[329] The flower-inducing principle was identified as salicylic acid.[330] Numerous attempts to induce flower initiation in *Xanthium* with this chemical given either alone or in combination with GA_3 and/or kinetin have yielded negative results.[330] This fact, together with the occurrence of salicylic acid in both the flowering and the vegetative honeydew makes it unlikely that this compound is the floral hormone.

The outcome of all this work is really discouraging. The lack of success could be due either to absence of a reliable bioassay of florigen or to our inability to extract this hormone.

a. *Bioassays of the Floral Hormone*

The simplest assays are no doubt intact plants kept in strictly noninductive conditions. In order to avoid translocation problems with these plants, it is advisable to apply the extract directly on the receptor bud or to a leaf which has been demonstrated to supply this meristem with assimilates. A difficutly arises, however, since one may suspect that leaves in noninductive conditions produce high levels of endogenous floral inhibitors (see below) that could nullify the positive effect of any exogenously added flower-inducing substances. Removal of these leaves will not solve the problem because growth is often inhibited, thus preventing any production of new structures including floral ones.

A way to reduce inhibition is to grow the test plants under a photoperiodic regime allowing marginal flowering, but the results are then bound to be inconclusive and it is known that a large array of exogenous substances can increase the floral response in these conditions (Volume II, Chapter 6).

Another way to avoid leaf-generated inhibitors is to use excised stem apices grown in vitro. This method has the further advantage of reducing in the plant material the proportion of tissues, such as mature leaves, stem, and roots, that is supposed not to respond to the flower-promoting principle, leaving only the target meristem and few very young leaves. It is, however, not without difficulties. Excised stem tips of some species, e.g., the SDP *Perilla,*[331] may flower almost automatically provided they are stripped of their unfolded leaves, precluding their use as an assay of florigen. Finally, in order to get some growth in excised shoot apices, it is often necessary to add various growth regulators to the culture medium; these may also be factors required for floral evocation, but there is no way to distinguish between the two possibilities.

Other convenient assays are the photoperiodic duckweeds grown sterily in vitro. However, there is the complication with these plants that they respond to a wide variety of seemingly unrelated chemicals (Volume I, Chapter 2 and Volume II, Chapter 6). The significance of these responses is far from understood. Clearly, there is no ideal bioassay of florigen and it is not easy to see how to develop one.[92]

b. Inappropriate Extraction Procedures

Imperfect bioassays alone might not be responsible for our failures in identifying the floral hormone. Another reason, *perhaps the major one,* is the use of inappropriate extraction techniques. The hormone may be (1) insoluble in methanol or acetone or (2) a high molecular weight compound denaturated in these solvents. The last possibility was suggested many years ago by Bonner and Liverman, based on the infectious nature of induction in *Xanthium* (see Volume I, Chapter 3, Section VII).[95]

On the other hand, if the floral stimulus is not one, but several different compounds acting in sequence or in concert, but at specific concentrations, only rarely and even then quite fortuitously would extracts and reapplications of one or more fractions match the phloem mobile components. Thus, our classical procedures of extraction and fractionation may be ineffective and the physiological evidence at hand does not offer a clue as to what kind of compound(s) to search for. The identification of florigen remains essentially an exercise in the exploitation of chance, admittedly a very uncomfortable situation for scientists.

2. Applications of Known Chemicals to Vegetative Plants

An empirical approach to the question of florigen identity consists in treating vegetative plants by a variety of known chemicals; any substance that happens to be on the laboratory bench the day of the experiment will do, including very sophisticated molecules such as cAMP, prostaglandins, DNA, etc. Behind this effort, is the hope of finding at least one compound which exhibits universal florigenic activity. Despite the huge amount of work of this type (to be reviewed in Volume II, Chapter 6), the outcome is so far inconclusive in the sense that no compound has been found which is effective in *all* higher plants.

Many unrelated chemicals produce a clear florigenic effect in a *limited number* of species. Preeminent among these are the GAs, but also carbohydrates and various growth regulators, such as the auxins, cytokinins, and ethylene. However, because these compounds are not universal promoters of flower initiation, proponents of the florigen theory suggest that their action is nonspecific or pharmacological and by definition they must be dismissed from the list of candidates for the floral hormone.

C. Modified Florigen Theories

The difficulties in reconciling all data into the simple florigen concept was a prod to its proponents for further elaboration.

Chailakhyan himself enunciated a modified florigen theory based on the discovery by Lang in 1956 that GA_3 is able to cause flowering in several LDP and some LSDP grown in SD.[332] This compound appeared thus to substitute for the LD requirement in these plants. The effect of GA_3 is, as a rule, most marked in species having the rosette habit of growth in the vegetative state. One of the most spectacular results of treatment of rosettes with this growth regulator is a tremendous stimulation of stem elongation (bolting), a process usually closely associated with the onset of flower formation in rosette plants. Another important observation at that time was the inability of GA_3 in causing flowering of SDP in noninductive conditions. In an attempt to account for the role of GAs in flowering, Chailakhyan proposed in 1958 that the leaf-generated floral hormone is composed of two complementary substances: one belonging to the family of GAs and the other to the "anthesins", a group of still undiscovered compounds.[333] The GA would be the limiting factor in *all photoperiodic plants* grown in SD and anthesin the limiting factor in LD conditions. This theory has never been widely accepted because its experimental support is very poor. Indeed, applications of GA_3 to LDP, LSDP, or SLDP grown in SD fail in many cases to promote flower initiation, and may even inhibit flowering in as many other species representing all response types (Volume II, Chapter 6). Also, GA_3 was found to substitute for SD in some SDP. Contrary to one prediction of the anthesin theory, graft combinations between a vegetative SDP, grown in LD and thus presumably rich in GAs, and a vegetative LDP, grown in SD and presumably rich in anthesins, do not generally result in flowering. One positive case only is known (Table 4). Despite this, the anthesin hypothesis must be regarded as an important advance in concept because it is the first that (1) implies a *complex* floral hormone and (2) suggests that different limiting factors may exist in different groups of plants.

Carr considered that we really have no evidence that the substance which is transmitted through a graft union is the same as that which is exported out of a leaf immediately after induction.[327] On this basis, he proposed another modified florigen theory in which the immediate product of leaf photoinduction (primary induction) is labile and diffusible, equivalent to "florigen". He reserved the term "floral hormone" for the stable substance, transmissible by grafting, which is produced in induced plants only after the occurrence of a permanent endogenous change (secondary induction). Although there is no evidence whatever against this interesting idea, there is also at present no experimental evidence in its favor.[70] Thus, even the latest refinements of the florigen theory are not satisfactory.

CONCLUSIONS

After more than 40 years of extensive and careful work no isolation of the presumptive universal floral hormone or florigen has yet been made. This failure could be due not only to absence of a reliable bioassay, but also, and more probably, to use of inappropriate extraction techniques.

The evidence at hand offers no clue as to what kind of compound(s) to search for.

Clearly, a transmissible flower-promoting material is present in many plants, as shown in grafting experiments, but whether this material is simple or complex, specific or unspecific was not (and cannot be) answered in these experiments. In view of the many failures and anomalies encountered in the attempts to transmit the floral stimulus through a graft union, it appears that the results may be equally well interpreted in terms of a universal floral hormone, the transmission of which exhibits many irregularities, or in terms of several different floral hormones each acting only in certain species.

The experimental evidence indicates that the floral stimulus(i) generated in induced leaves of photoperiodic species is generally, although not universally, transported to receptor meristems in the phloem along with the assimilates. Sequential defoliations in single-cycle plants suggest that this stimulus(i) moves out of the leaves soon after it is formed and is translocated for a short time following induction.

Despite all the difficulties mentioned above, the concept of florigen has dominated the field of flower initiation since 1937. How do we account for its continuing attractiveness? Simplicity of this theory is certainly part of the answer. Another reason is no doubt the great influence played by the discovery early in this century of circulating sex hormones in animals and the conceptual framework that developed in developmental zoology following this discovery. This tendency to compare plant flowering and animal sexuality was fairly well marked in early days, as evidenced by the frequent use of the word "sexual" in the literature on flowering physiology. So far, however, no specific organ-forming hormone has been identified in higher plants and the known plant growth substances, sometimes called "plant hormones", differ markedly from animal hormones in a number of ways. Perhaps the most important differences are that they are usually not produced in specific organs and have many different effects upon various plant tissues and processes, depending on other parameters.

III. THE THEORY OF FLORAL INHIBITORS

A counter theory to that of the floral hormone was postulated by Lona, Von Denffer, and others in the years 1949 to 1950.[334,335] This concept supposes that plants grown in conditions unfavorable for flowering produce one or several floral inhibitors and that flower initiation occurs in conditions preventing the production of these inhibitory compounds. Thus, induction lowers the concentration of the inhibitor(s) below some threshold.

The inhibitor concept started from the early observation that (1) flower formation can be obtained in the LDP *Hyoscyamus* and the SDP *Chenopodium amaranticolor,* kept in strictly noninductive photoperiodic conditions, simply by continuous removal of all leaves, combined with sugar feeding in the case of *Chenopodium,*[296] and (2) noninduced leaves, positioned so that they do not presumably interfere with translocation of the floral stimulus, have an inhibitory effect in spinach.[336]

A. Experimental Evidence Concerning Floral Inhibitors

1. Grafting Experiments

In many grafting experiments, there is no evidence for a floral inhibitor; that is the vegetative partner does not seem to inhibit or decrease the flowering of the reproductive partner. However, this observation might not be of general significance since (1) the vegetative receptor plants are usually defoliated in these experiments and (2) the inhibitor is supposed to be generated in the leaves.

On the other hand, graft combinations between an early and a late pea variety by Paton and Barber show that the early variety Massey flowers later when grafted on the late variety Telephone than when it is grafted on its own stock (Figure 3). Thus,

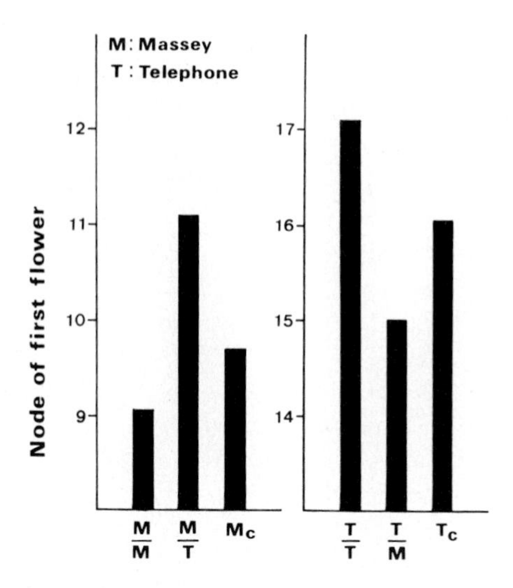

FIGURE 3. Influence of control grafting (M/M and T/T), experimental grafting (M/T and T/M), and removal of cotyledons (M_c and T_c) on node of first flower of scions of the early M variety (left) and the late T variety (right) of *Pisum sativum.* (Adapted from Paton, D. M. and Barber, H. N., *Aust. J. Biol. Sci.*, 8, 230, 1955.)

the cotyledons and root system of a late Telephone stock synthesize substance(s) that delays the response of meristematic tissues of an early Massey scion to floral promoters. In reciprocal grafts, a scion of Telephone flowers at an earlier (lower) node under the influence of a Massey stock (Figure 3). Cotyledon excision causes a slight increase in the number of nodes preceding the first flower and thus slightly inhibits flower initiation in early Massey peas, whereas the same treatment lowers the position of the first flower in late Telephone peas (Figure 3). These results suggest that earliness in pea is controlled by a balance between transmissible flower-inhibiting and flower-promoting materials.[21,337-339] Both materials are produced in the cotyledons and presumably also in the leaves. Murfet and Reid have shown that a specific gene (gene *Sn*) in late peas is responsible for the production of the graft-transmissible inhibitor in the cotyledons and the shoot. The activity of this gene is high in SD, but suppressed under LD or low-temperature conditions.[14,340]

Recently, Lang and associates observed that a noninduced scion of the LDP *Nicotiana sylvestris* greatly delays or completely suppresses flowering of a day-neutral tobacco cultivar (cv. Trapezond) used as stock (Figure 4 right). Also, the growth habit of the Trapezond stock becomes dwarf-like as seen in Figure 4. These results strongly indicate that noninduced *N. sylvestris* plants produce a transmissible flower-inhibitory and growth-regulating material. Similar results have been obtained using noninduced annual *Hyoscyamus* (LDP) as scion, showing that the floral inhibitor of *N. sylvestris* is not unique.[341] This inhibitor is apparently not ubiquitous, however, since a scion from noninduced Maryland Mammoth tobacco (SDP) is relatively without influence on flower formation and growth in the Trapezond receptor.[341]

On the other hand, lack of an inhibitor does not seem to be typical for all SDP because Jacobs observed a flower inhibition in receptors of a day-neutral variety of *Coleus blumei* under the influence of grafted noninduced leaves of the SDP *C. fredericii.*[342]

Since the *Hyoscyamus* inhibitor is active in a tobacco receptor, one is tempted to generalize that this compound is the same in all Solanaceae, even all angiosperms.

FIGURE 4. Grafts of a scion of *Nicotiana sylvestris* (LDP) on Trapezond tobacco (DNP) used as stock. Left, LD conditions; right, SD conditions. The indicator shoot of the stock is in both cases on the left, arising above the graft region, the scion is on the right. The arrows indicate the graft regions. (From Lang, A., Chailakhyan, M. Kh., and Frolova, I. A., *Proc. Natl. Acad. Sci. U.S.A.*, 74, 2412, 1977. With permission.)

Fortunately, we have the "florigen" theory as an excellent example of how misleading a single factor hypothesis can be and we are not likely to make the same mistake again.

2. Parasitic Plants

We return now to the experiments with dodder on soybean in which transmission of a floral promoter cannot be detected (Section II. A. 2 in this chapter). Fratianne further observed that dodder does not flower on intact soybean maintained vegetative in noninductive LD, but flowers following simple removal of all leaves on these soybean plants.[306] Soybean plants kept in inductive SD linked with dodder bridges to soybean plants kept in noninductive LD reveal that: (1) the induced partner is unable to cause flower initiation in the LD partner and (2) the LD partner has a delaying effect on the flowering of its SD mate. The most plausible explanation for these results is that a transmissible floral inhibitor is produced by noninduced leaves of soybean.

3. Studies on the Transport of the Floral Inhibitory Material in a Single Plant

The strawberry, *Fragaria ananassa,* is a useful material to investigate translocation of substances because of its habit of forming daughter plants connected to parent plants by stolons. Guttridge fully exploited this advantage by using a facultative SD variety of this species and pairs of runner plants joined by the stolon, one partner of each pair being grown in SD and the other in continuous light or in SD with night interrupted by a light break.[343,344] He found that flower formation in the SD partner is much delayed under the influence of the LD mate. Flowering is inhibited, but vegetative growth, i.e., petiole length, leaf area, and stolon production is promoted. All these effects are proportional to the number of leaves retained on the LD partner. Attempts to demonstrate the transmission of a floral promoter from the SD partner to the LD one have given negative results, even when conditions were created which presumably favor transport of assimilates from the former to the latter. The results are best interpreted by assuming that flower initiation is negatively controlled in this plant: the inductive effect of SD is essentially the arrest of production of a flower-inhibitory, growth-regulating hormone. Support for this interpretation comes from the fact that defoliated individuals of this strawberry variety flower in LD while intact plants fail to do so.[345]

A floral inhibitor has been also detected by Evans in the LDP *Lolium temulentum*.[346] The floral response of this plant to a single LD depends on the balance between the SD and the LD leaf areas. Ten cm^2 of a sensitive LD leaf are sufficient for inflorescence initiation in the absence of SD leaves, but are insufficient when five leaves are in SD. These noninduced leaves are located below the induced one, and as shown by $^{14}CO_2$ labeling, they are not sinks for assimilates from the induced upper leaf nor do they reduce translocation from the upper LD leaf to the shoot apex.[347] We are thus forced to conclude that a transmissible floral inhibitor is produced by the noninduced leaves in *Lolium*. These leaves supply little assimilates to the shoot apex suggesting that the inhibitor is either not translocated in the assimilate stream or is transported along with the assimilates, but then is active at very low levels or produced in high amounts by noninduced leaves. Removal of noninduced leaves in different groups of *Lolium* plants at various time intervals following the start of exposure of one upper leaf to a single LD reveals that the longer the noninduced leaves remain on the plant, the lower is the proportion of plants subsequently initiating inflorescences (Figure 5). Inflorescence initiation is almost totally suppressed when the noninduced leaves are kept on the plant for 32 hr after the start of the LD.

In *Rottboellia exaltata,* a SDP requiring six SD for induction, a single LD given to the lower leaves while the uppermost leaf blade is exposed to the six SD *accelerates* inflorescence development if given early in the inductive sequence and *inhibits* it if given late in this sequence. Again, the relative position of the leaves eliminates the possibility of noninduced leaves altering movement of assimilates (and, hence, the floral stimulus) from the upper induced leaf to the shoot apex and implies most probably the formation of translocatable promoters and inhibitors in noninduced leaves.[74]

In *Salvia occidentalis,* another SDP, with two leaf pairs, when the upper leaf pair is in continuous SD and the lower pair receives 3 days of continuous light after ten SD (Volume I, Chapter 3, Figure 4), the appearance of flower buds is markedly delayed in comparison with plants in which both leaf pairs are subjected to SD only. This delay is observed regardless of whether the lower leaf pair is removed immediately following

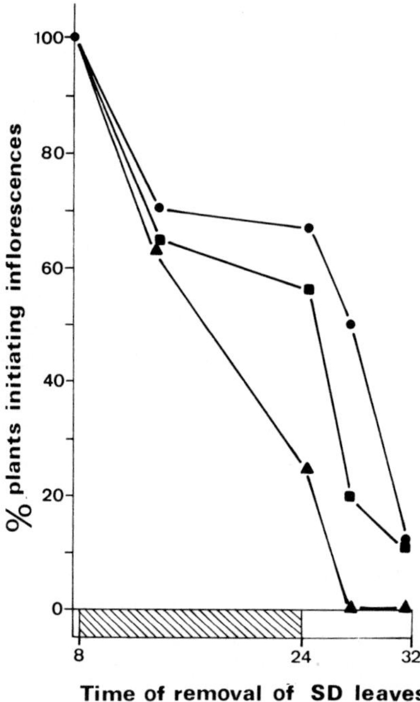

FIGURE 5. Inflorescence initiation in the LDP *Lolium temulentum* as a function of time of removal of lower leaf blades kept in SD conditions, during exposure of the sixth leaf (counting acropetally) to a single LD. LD leaf area: 26.6 cm^2; SD leaf areas 21.5 cm^2 (leaf 5, ●); 45.9 cm^2 (leaves 3, 4, and 5, ■); or 172.4 cm^2 (leaves 1 to 5, ▲). The hatched area indicates the period when leaf 6 was illuminated while the lower leaves were in darkness. (From Evans, L. T., *Aust. J. Biol. Sci.*, 13, 429, 1960. With permission.)

the 3 days of continuous light or left indefinitely on the plants.[83] A very small area of leaf tissue in continuous light is sufficient to secure a substantial response, indicating that the transmissible inhibitory material is either actively synthesized by the noninduced leaf cells or is active at very low levels.

In the SDP *Xanthium*, the story concerning inhibitors is quite confused. We have seen above that mature noninduced leaves located between the induced leaf and the receptor meristem are inhibitory because they interfere with the movement of assimilates and accompanying floral stimulus from the induced leaf.[320] This effect does not preclude, however, other inhibitory actions of these leaves. Indeed, according to Gibby and Salisbury, and to Lincoln and co-workers, noninduced mature leaf tissues have also an *active* inhibitory action.[46,315] This action is a true photoperiodic effect in the sense that it is a property not only of leaves exposed to LD, but also of those receiving SD with a light break in the dark period (Figure 6). Apparently, when a *Xanthium* leaf is not producing the floral stimulus under the effect of inductive conditions, it inhibits flowering under the effect of noninductive conditions.

Searle finds also that lower leaves maintained in noninductive conditions (continuous light) may affect the floral response from a single upper induced leaf in *Xanthium*.[348] These leaves have a promotive effect if present for 7 to 25 hr after end of

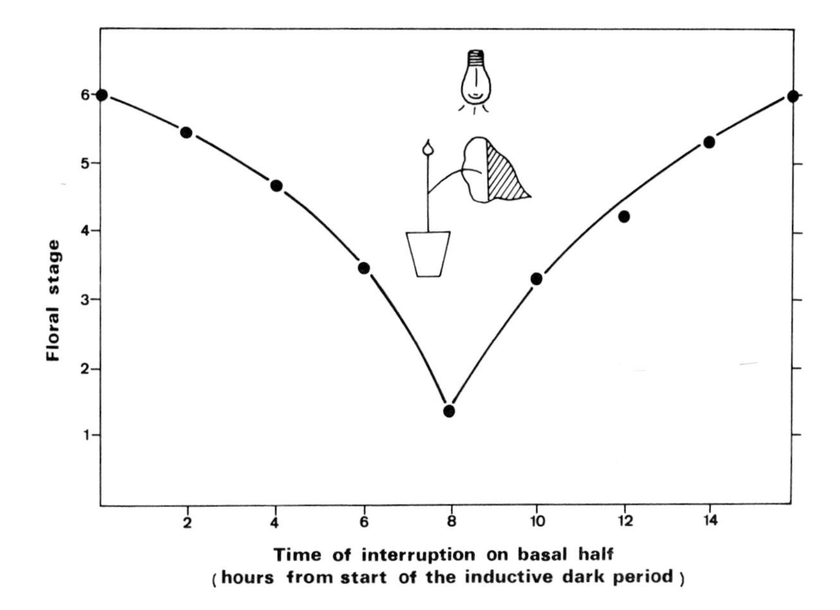

FIGURE 6. Inhibition of flower initiation in unifoliate *Xanthium strumarium* by a 5-min light interruption given at various times during an inductive 16-hr dark period to the basal half of the leaf. The 16-hr dark period on the tip half was uninterrupted. (From Gibby, D. D. and Salisbury, F. B., *Plant Physiol.*, 47, 784, 1971. With permission.)

the inductive dark period, but beyond this time they inhibit, a result reminiscent of that for *Rottboellia.* Note that in this case, as in *Rottboellia,* the noninduced leaves are not inserted between the induced leaf and the target meristem. Searle's interpretation is that transmissible substances produced in continuous light are responsible for these effects. Gibby and Salisbury conclude, on the contrary, that the LD-inhibition in *Xanthium* is a localized (immobile) condition.[46] Thus, the nature of this inhibition is not at all clear.

Taken in its entirety, the experimental evidence favors the existence of transmissible floral inhibitors, presumably acting at the meristem, rather than the existence of immobile inhibitors interfering within the leaves with the production, loading or export in the phloem of the floral promoters. This evidence is so far limited to a small number of photoperiodic species and whether floral inhibitors exist in all species remains to be determined. Results of many experiments may be entirely explained without the assumption of an inhibitor. It would be premature, however, to conclude from this that most plants are devoid of inhibitors because of the difficulties in establishing their presence. This point is best exemplified by the case of the LDP *Hyoscyamus.* If only six leaves remain on the plants, one in LD and five in SD, flowering occurs as rapidly as if the SD leaves were removed.[349] This result suggests that no inhibitor is produced by SD leaves whereas, on the contrary, grafting experiments described above clearly reveal that noninduced *Hyoscyamus* leaves release an inhibitor.

Floral inhibitors, whatever their nature, do not accumulate in photoperiodic species. This is inferred from the fact that these plants as a rule require exposure to progressively fewer inductive cycles as they age (Volume I, Chapter 7). Noninductive conditions preceding induction have no clear cumulative effect. In experiments on fractional induction in SDP, Schwabe and others have observed that the inhibitory effect of interpolated LD given consecutively is not cumulative beyond a certain limit.[87] This suggests again that the eventual inhibitor is not accumulated.

4. Cultures of Isolated Apices

Evidence for an inhibitor comes also from the experiments of Raghavan and Jacobs with excised apical buds of the SDP *Perilla.*[331] These apices cultured in vitro can initiate a normal inflorescence in SD and a rudimentary* one in LD, indicating that these apices have an inherent capacity to flower regardless of the photoperiodic regime. If the apices have two pairs of unfolded leaves, they behave as intact plants flowering only in SD. These unfolded leaves when implanted in the same medium, but separated from the apices also inhibit flower initiation irrespective of the daylength regime. This observation has led Raghavan and Jacobs to propose that young leaves of *Perilla* produce and release a floral inhibitor.

Excised apices of the LDP *Sinapis* behave similarly, initiating flower primordia in noninductive as well as in inductive conditions.[32]

B. Identification of Floral Inhibitors

The identification of inhibitors requires the use of sensitive and specific bioassays. Floral inhibitors, in order to be considered specific, cannot at the same time inhibit general meristematic functions. Schwabe used *Kalanchoe* plants partially induced by 9 to 15 SD cycles.[350] Only one young mature leaf is subjected to the inductive regime and the test extract is injected into that leaf two or three times during induction. Axenic cultures of partially induced duckweeds and excised shoot apices from partially induced plants of *Viscaria* (= *Silene candida*) have also been used to test extracts added to the culture medium.[329,351] With both the *Kalanchoe* and the *Viscaria* assays, the crude sap and the aqueous extract of noninduced leaves of *Kalanchoe* were found to contain a potent floral inhibitor. This material is absent in induced leaves of the same species.[350,351] Honeydew collected from aphids feeding on vegetative or flowering *Xanthium* has been analyzed for presence of floral inhibitors.[329] Several fractions of both the vegetative and the flowering honeydew have an inhibitory effect on flower initiation in the duckweed bioassay.

The chemical nature of these presumptive inhibitors is entirely unknown. A claim that gallic acid is the inhibitory compound produced by *Kalanchoe* in noninductive LD has not been verified in further study.[92] Similarly, an early suggestion that ABA is the SD inhibitor in the LDP *Lolium* has not received support in subsequent work (Volume II, Chapters 6 and 7).

C. Balance Between Inhibitors and Promoters of Flower Initiation

It is of interest that most species for which floral inhibitors have been proposed, e.g., *Nicotiana sylvestris,* annual *Hyoscyamus,* peas, *Lolium, Rottboellia, Xanthium,* and *Perilla,* also produce translocatable flower-promoting materials. Consequently, in the views of many researchers, like Evans, Wareing and El-Antably, Reid and Murfet, etc., floral evocation at the meristem is controlled by a balance between these two kinds of compounds.[15,339,352]

The idea that flower initiation is controlled by a change in the ratio of two or more flower-promoting and flower-inhibiting substances is challenged by Lang and by Zeevaart on the basis that it is difficult to see how such a ratio: (1) moves over long distances from leaves to meristems without change, (2) is maintained in the meristem by a single induced leaf in the presence of several noninduced ones that are presumably exporting an unfavorable ratio, and (3) survives a period of noninductive conditions (in cases of successful fractional induction) which results in an unfavorable ratio.[21,320] These difficulties can be met by assuming that the favorable ratio must be produced *only at the meristem and only during a limited time interval.* The results with *Rott-*

* In LD *Perilla* apices produce inflorescences whose florets are totally devoid of fertile tissues.

boellia and *Xanthium,* quoted above, further suggest that substances produced by noninduced leaves, while usually inhibitory to inflorescence initiation, may *promote* early phases of evocation. Thus, the most favorable ratio between subtances from induced and noninduced leaves probably changes as evocation proceeds.

CONCLUSIONS

There is increasing experimental evidence favoring the existence of transmissible floral inhibitors, at least in some species. These inhibitors are apparently produced in leaves exposed to photoperiodic conditions unfavorable for flower initiation and presumably act at the shoot meristems. These inhibitory materials do not seem to accumulate in plants held in noninductive conditions. The evidence is so far purely of a physiological nature and the chemical nature of these inhibitors is entirely unknown.

In few species, both flower-inhibiting and flower-promoting substances have been detected. Flower formation in these cases seems to be controlled by a balance between the two classes of compounds.

With the kind of physiological work described in this chapter, even with the best of it, it is impossible to reach more conclusive statements about floral promoters and inhibitors. Results nearly always can be interpreted in several different ways and it is obvious that no satisfactory answers will arise from the continuation and refinement of the same type of experiments. On many occasions in this review we have seen that many basic ideas on which the field of flower initiation is founded are probably inadequate for large generalizations. Thus, if we want to do something more than refine concepts elaborated mainly in the 1930s and 1940s, it is urgent to rechallenge the infrastructure supporting them and hasten to explore new approaches to the problem. This is the aim of Volume II.

Chapter 7

AGE AND FLOWER INITIATION

TABLE OF CONTENTS

I. THE JUVENILE PHASE

Regardless of the conditions in which they are grown, most plants grow vegetatively for some time after sowing. This is the "juvenile" or "maturation" phase. So long as a plant is in this phase, vegetative growth proceeds and the plant is totally insensitive to conditions which later promote the floral transition. A typical juvenile phase, measured in years, is common in woody perennials, but is also found in some herbaceous annuals or biennials where its duration may vary from a few days to several months. Juvenility, i.e., a delay in the onset of reproductive development, is part of an evolutionary strategy where increase in size is a key to success. Thus, depending on the ecological "niches" to which they are adapted plants may flower extremely rapidly at a very small size or may do so only after a protracted period of vegetative growth and attainment of a minimal size. At the completion of this juvenile phase, the plants are said to be adult or to have reached the condition of "ripeness-to-flower", i.e., they have become sensitive to conditions which eventually cause flower initiation.

For the majority of herbaceous photoperiodic species, there is clearly no period of total inability to flower, but the sensitivity to daylength increases with increasing age of the plants. In *Sinapis,* for example, the number of LD required for induction is 6 to 7 at 15 days, 2 at 30 days, and 1 at 60 days from sowing.[353] A similar decrease in the minimum number of inductive cycles with age has been found in Biloxi soybean, *Kalanchoe, Perilla,* and *Chenopodium amaranticolor,* annual *Hyoscyamus, Lolium, Silene armeria,* etc.[21,53,76,334] Even *Lemna paucicostata* when grown from seeds, admittedly an unusual way to cultivate this plant, exhibits an early phase of weak sensitivity to SD.[354] Although these plants are clearly devoid of a true juvenile phase, it is a common practice to call plants juvenile during their early period of growth when they exhibit a poor photoperiodic response.

Some photoperiodic species, as the LDP *Fragaria vesca semperflorens,* the SLDP *Campanula medium,* and the LSDP *Bryophyllum,* exhibit a typical juvenile phase which may extend over a considerable period of time.[93,113,317]

Full sensitivity to thermoinduction is generally attained at seed germination in cold-requiring annuals. After germination the sensitivity to cold may remain relatively constant as found in some winter cereals, e.g., rye and wheat (Figure 1, line A),[250,355] or first decrease and then increase as in the St race of *Arabidopsis* (Napp-Zinn),[356] *Cheiranthus allionii,*[240] and *Lactuca serriola* (Figure 1, line B).[357]

Because the mature embryo is already sensitive in these plants, response to low temperature must begin during embryogenesis. In several winter cereals, pea and other Leguminosae, the ripening embryo can be vernalized when still attached to the mother plant or when excised, developing ears are chilled.[21,275,294] Embryo sensitivity is highest at early developmental stages, e.g., 1 or 2 weeks after pollination, and decreases at later stages. Ripening embryos of *Arabidopsis* are apparently not sensitive to cold.[256]

Most cold-requiring plants of the biennial or perennial types do not respond to a chilling treatment at the seed stage or during germination, i.e., they possess a juvenile phase (Figure 1, line C). Examples of this are biennial *Hyoscyamus,*[245] *Lunaria annua,*[290] *Dianthus barbatus,*[271] *Oenothera biennis,*[241] *Cardamine pratensis,*[276] *Geum urbanum,*[31] and several species exhibiting a direct response to chilling, namely Brussels sprouts, stocks, and cabbage.[21,255] When the juvenile phase is over, the sensitivity to cold increases more or less rapidly to a maximum which may be maintained for an extended period of time (Figure 1, line C; see also Volume I, Chapter 5, Table 4).

In the temperate zone, the biennial habit of many spring-planted, cold-requiring species is primarily related to the existence of this juvenile phase which prevents thermoinduction during the cool days and nights of the first spring.

In few biennials, as carrot, red garden beet, turnip, *Cichorium intybus,* and the

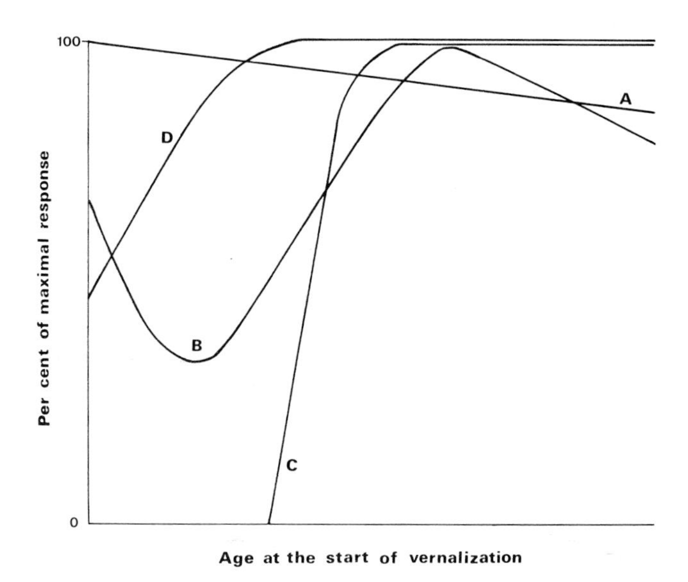

FIGURE 1. Diagram showing schematically response to vernalization as a function of age in *Secale cereale* (Petkus winter rye) (A), St race of *Arabidopsis thaliana* (B), biennial *Hyoscyamus niger* (C), and *Beta vulgaris* var. Egyptian Flat Round (red garden beet) (D). The time scale in abscissa is different for the different species. (Adapted from Purvis, O. N., *Encyclopedia of Plant Physiology,* Vol. 16, Ruhland, W., Ed., Springer-Verlag, Berlin, 1961, 76.)

perennial grass *Lolium perenne,* the germinating seeds can be vernalized, but the response is usually far less at the seed stage than at subsequent developmental stages (Figure 1, line D). According to Lang, seed thermoinduction is in principle impossible in biennials but, if the plants are able to make some growth during the chilling treatment, then they may reach the earliest stage of sensitivity.[21]

Many so-called absolute photoperiodic and cold-requiring plants may ultimately flower in continuous ''noninductive'' conditions, indicating that they are not in a complete steady vegetative state in these conditions, but are progressing slowly towards the reproductive state. Clearly, the difference between these plants and those having a quantitative response is thus simply a question of degree. Claims of maintaining the vegetative state for years have been made for a few species, but in our experience this has been impossible in conditions favoring vigorous growth *unless* steps are taken to remove the older mature shoots continuously. This includes experience with the LDP *Fuchsia* and *Sinapis* and the SDP *Xanthium.* Other species in which older plants have also been seen bearing flowers in ''noninductive'' conditions are the SDP *Pharbitis,*[59] *Kalanchoe,*[358] and soybean,[21] and the cold-requiring *Geum urbanum*[112] and *Lunaria annua.*[104]

II. MINIMAL LEAF NUMBER

In herbaceous species, the number of nodes to first flower may be used as a measure of the length of the juvenile phase. Indeed, the shoot apical meristem invariably produces some leafy nodes below the earliest flower or inflorescence. This observation has led to the concept of ''minimal leaf number'', the irreducible vegetative growth produced prior to flower initiation in plants held in conditions optimal for flowering from the very beginning of germination. Efforts have been made to determine this number in some plants, but it is difficult to be absolutely sure that the values that have

Table 1
MINIMAL LEAF NUMBERS IN *SECALE CEREALE* (PETKUS WINTER RYE)

Type of embryo	Leaves present in embryo	Leaves added after imbibition	Minimal leaf number
Normal	3—4	3—4	7
Dwarf[a]	2.7	3	5.7

[a] Harvested prematurely from nitrogen-starved plants.

Adapted from Gott, M. B., Gregory, F. G., and Purvis, O. N., *Ann. Bot.* (London), 19, 87, 1955.

Table 2
INHERITANCE OF EARLINESS TO FLOWER IN THE SDP *CHRYSANTHEMUM MORIFOLIUM* AND THE LDP *SILENE ARMERIA*[363,364]

Species	Cultivar or line	Number of leaves and bracts below the capitulum in LD
Chrysanthemum morifolum	Tuneful	57.8
	Polaris	39.7
	Pollyanne	29.8
	Bright Golden Anne	18.1
		Days from sowing to visible flower buds in LD
Silene armeria	S4.1	40
	S1.1	48
	S1.4	99

been obtained are really irreducible minima.[359] Usually the lowest leaf number is produced when the plant is not only kept under optimal conditions for flowering, but also under extreme nutrient starvation. Minimal leaf numbers in some very sensitive photoperiodic and cold-requiring plants are often close to the number of leaf primordia present in the ripe embryos. Invariably, however, a few additional leaves at least are added between the start of germination and initiation of the first flower. In Petkus winter rye, normal and dwarf embryos produce the same number of leaves between imbibition and the floral transition (Table 1).[29] The physiological significance of these observations is unclear, i.e., there is no evidence that the number of leaves in itself is a determinant of flower initiation.

Flower or inflorescence primordia have apparently been found in ripe embryos of some varieties of maize and peanut.[360,361] Note that the embryos of these species are known to initiate quite a number of leaves in addition to the cotyledons during seed maturation.

III. GENETICAL BASIS FOR JUVENILITY

In most annual species, strains which flower far earlier than others are known. Quite generally, late strains or varieties require more inductive cycles and/or have a longer juvenile phase than early ones.[21,362] The question of "earliness" is thus obviously related to, perhaps inseparable from, the question of juvenility. Earliness is an inherited character as shown in Table 2 for *Chrysanthemum* and *Silene*. In *Chrysanthemum*, a

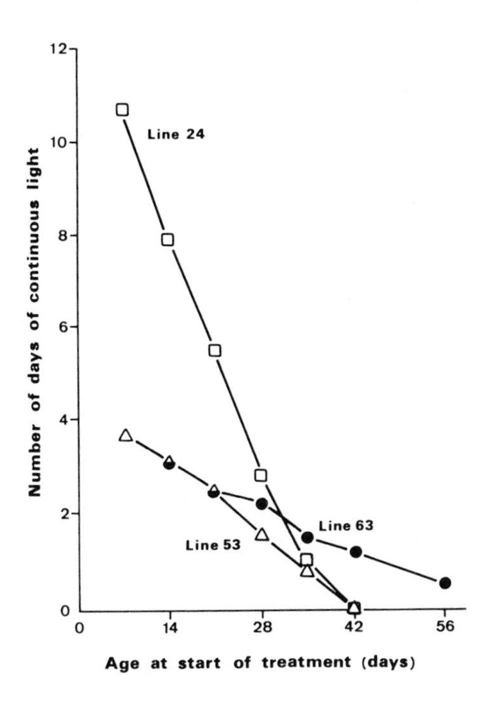

FIGURE 2. The effect of age on the number of days of continuous light required to induce 50% flower initiation in lines 53 *(lf e Sn hr)*, 63 *(lf e Sn Hr)*, and 24 *(Lf e Sn hr)* of *Pisum sativum*. The plants are grown in 8-hr SD before and after treatment. (From Reid, J. B. and Murfet, I. C., *J. Exp. Bot.*, 28, 811, 1977. With permission.)

facultative SDP, grown in LD, the numbers of leaves and bracts inserted on the main stem below the involucral bracts of the terminal capitulum are very different in different cultivars.[363] These numbers may vary between experiments, but when the cultivars are ranked in order of leaf and bract number, their relative positions are similar in each experiment.

In *Silene*, a number of lines differing by their degree of earliness were isolated by Wellensiek from an unselected population. Earliness in this case depends on the presence of the dominant gene *E*, lateness resulting of the presence of *e*.[364]

The numerous varieties of *Pisum sativum* (garden pea) can be divided by their habit of flowering into several types, ranging from the early day-neutral to the late LD type. The genetic analysis of Murfet has revealed that four major loci, *lf, e, sn*, and *hr*, are responsible for the gross differences in flowering behavior.[14] Genotype *lf e sn hr* is an early flowering DNP. Gene *Sn* confers an increase in the flowering node and genotype *lf e Sn hr* (line 53) is a late flowering, quantitative LDP.

The flowering node of peas appears to be determined by the time at which the ratio of a floral promoter to a floral inhibitor rises above a certain threshold level (Volume I, Chapter 6). The physiological evidence indicates that gene *Sn* controls the production of this graft-transmissible inhibitor.[337,340] Three pure lines, all containing gene *Sn*, require fewer inductive cycles as they age (Figure 2), suggesting that the activity of this gene decreases as the plant ages. However, the curves for lines 24, 53, and 63 differ and, according to Reid and Murfet, these differences largely reflect the action of genes *Lf* and *Hr*.[339] The action of gene *Lf* is indicated by comparing the curves for lines 53 *(lf)* and 24 *(Lf)*. Gene *Lf* does increase the number of inductive cycles required to induce flower initiation in young individuals of line 24. Thus, the *lf* locus influences the minimum flowering node, i.e., the juvenile phase. The effect of gene *Hr* can be seen by comparing the curves for lines 53 *(hr)* and 63 *(Hr)*. There appears that *Hr* has no effect during the first 3 weeks of growth, but from this time onwards it increases the number of inductive cycles required to flower. Thus, *Hr* seems to reduce the effect of age on the gene *Sn*. In other words, it prolongs the vegetative phase so that the balance between promoter and inhibitor approaches the critical threshold slowly in genotype *lf e Sn Hr* (line 63), which is a near obligate LDP.

Earliness is also an inherited trait in many other species, including cereals such as rice,[365] and woody perennials such as *Pyrus* (pear) and *Malus* (apple) where large differences in age to first flower are found in seedlings produced from early and late parents.[366]

IV. ACCOUNTING FOR THE JUVENILE PHASE

There may be many reasons why a young plant is unable to flower even when subjected to otherwise favorable conditions. Complete discrimination between the operation of these factors in some experiments is difficult because they may be confounded. The five most obvious reasons are now discussed.

A. Insufficient Leaf Area

This possibility is often disregarded since it is known that very little leaf tissue is sufficient in several photoperiodic species to secure a substantial flowering response. However, insufficient leaf area has been found to reduce or delay flower initiation in *Pharbitis* (Volume I, Chapter 3, Table 3) and other species, and it is clear from in vitro experiments that very small leaf area is sufficient for photoinduction only when carbohydrates are present in the medium (Volume I, Chapter 3).

That the photosynthetic contribution of leaves to attainment of ripeness-to-flower cannot be ignored is further suggested by the observation that good light conditions that permit high photosynthetic rates, that is high photon flux density and/or LD, during the early stages of growth shorten the juvenile phase in many species. This has been observed, for instance, in the LDP *Silene armeria*,[367] the SLDP *Campanula medium*,[113] the cold-requiring plants *Oenothera biennis*,[241] *Lunaria annua*,[367] *Cheiranthus allionii*,[240] *Cardamine pratensis*,[276] and in several woody perennials.

Also, the increase in thermoinductibility with age in some biennials and perennials goes parallel to an increase in leaf number and area, and leaf removal may abolish the sensitivity to cold in many of them.[31,240,241,260,265] This requirement for leaves, which is not found in celery and adult carrot plants,[242,368] is probably related to the provision of food materials by these organs to the shoot apex. The shortening of the juvenile phase that is observed in good light conditions could be similarly interpreted in terms of a more rapid increase in leaf number and area in these conditions.

B. Unfavorable Ratio of Immature to Mature Leaves

Some workers believe that flower initiation cannot occur before a definite ratio of immature- to mature-leaf area is reached. Evidence concerning this possibility has been found in several species, e.g., the SDP soybean,[369] the LDP *Scrofularia arguta*,[370] and the DNP tomato,[371] in which removal of young expanding leaves hastens flower formation in unfavorable conditions. Immature leaves may act by producing floral inhibitors or by interfering with the transmission of assimilates and accompanying floral stimuli from the lower mature leaves to the meristem (Volume I, Chapter 6). Young leaves are also the main tissues responsible for maintaining apical dominance and pro-

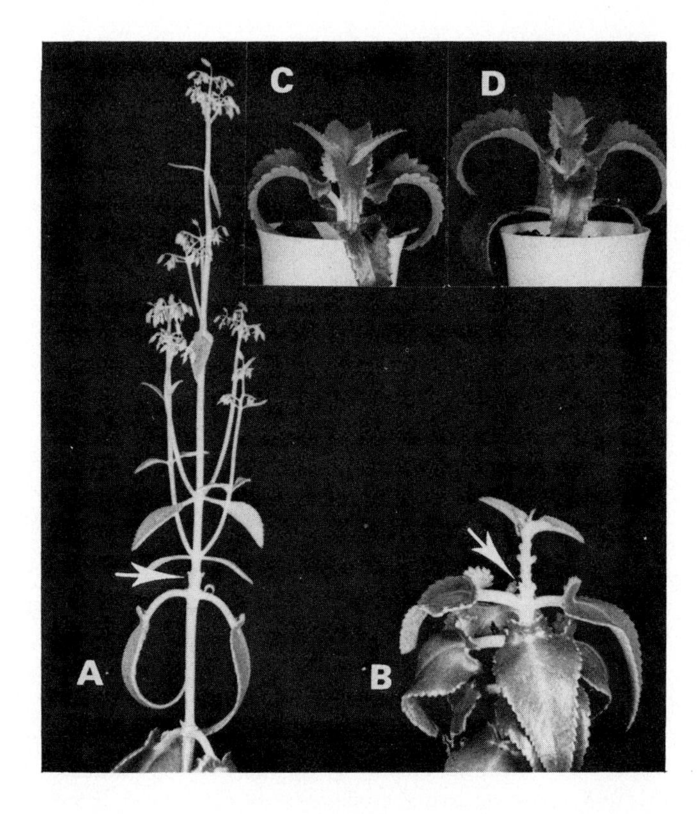

FIGURE 3. (A), A flowering juvenile *Bryophyllum daigremontianum* receptor scion in SD after grafting onto a donor stock in SD. The stock has been induced to flower by the change LD → SD and had received 43 SD when the grafts were made; (B), as a control, a juvenile scion in SD was grafted onto a vegetative stock in SD. Arrows indicate graft unions; and (C) and (D), juvenile ungrafted plants grown continuously in SD (C), or changed from LD to SD for 69 days (D). Both juvenile plants remain vegetative. (From Zeevaart, J. A. D., *Planta*, 58, 543, 1962. With permission.)

motion of flowering due to their removal might, in some species, result from the release of axillary meristems from correlative inhibition (Volume II, Chapter 1).

C. Leaf Insensitivity to Daylength

Insensitivity of the cotyledons and first-formed leaves to favorable daylength conditions is believed by most flowering physiologists to be the primary reason for juvenility in photoperiodic species. In *Xanthium*, for example, the cotyledons are incapable of responding to otherwise inductive SD.[372,373] In several other species, *Pharbitis*,[374] *Chenopodium*,[2] *Sinapis*,[8] *Brassica campestris*,[78] etc. however, cotyledons are fully inducible with the result that these plants may flower as young seedlings.

In the LSDP *Bryophyllum*, Zeevaart found that juvenile shoots flower rapidly when grafted onto mature flowering stocks, whereas ungrafted controls do not flower (Figure 3). This result led him to conclude that juvenility in this species is not a property of the meristem and is thus most probably related to an inability of leaves to respond to inductive photoperiodic conditions.[375] Examination of Figure 3, however, reveals that leaf area was much greater on the adult stocks (A) than on the juvenile plants (D). Thus, photosynthetic assimilate supply to the juvenile meristems was probably augmented when they were joined to the adult stocks. The control grafts between juvenile scions and noninduced mature stocks (Figure 3B) indicate that increased assimilate supply alone is not sufficient to cause flowering of juvenile scions, but it might be involved in their floral response when grafted on induced stocks. It remains true that the juvenile meristem responds to the floral stimulus, but the conclusion that juvenile leaves of *Bryophyllum* are incapable of producing this stimulus must be held in question. Subsequent work by Van de Pol has indicated that juvenile shoots of this species brought into flowering by grafting with mature donors are able in turn to serve as donors to other vegetative partners (indirect induction).[91] These juvenile shoots thus seem perfectly capable of reproducing the floral stimulus from an initial supply. The fact that applied GA_3 causes flowering in juvenile individuals held in SD suggests that juvenile leaves produce insufficient GA following the change from LD to SD.[317,376] Whether a high GA level is a prerequisite for leaf photoinduction or for apex evocation is discussed in Volume II, Chapter 6, Section IV. A. GA might also alter assimilate distribution so that more is available to the meristem than is the case without it.

In the LDP *Fragaria vesca semperflorens,* Sironval observed that daughter rosettes flower much earlier when attached to the flowering mother-plant than when detached.[314] The situation of these daughter rosettes is very similar to that of the juvenile scions of *Bryophyllum:* their meristems are responsive well before their leaves seem capable to produce the floral stimulus. However, as in *Bryophyllum,* an effect due to insufficient leaf area or improper ratio of immature to mature leaves in isolated daughter plants cannot be excluded.

Another Zeevaart's experiment with the SDP *Perilla* is equally instructive.[71] Two groups of plants were sown 4 weeks apart in LD in order to have the 2nd leaf pair of the second group and 5th leaf pair of the first group fully expanded at the same time. Leaves from these two nodes were then collected, reduced to a similar area, and grafted onto LD stocks. After a graft union was established, the donor leaves were exposed to a number of SD cycles. The results show that leaves of the 2nd node require about twice as many SD to become induced (as measured by the flowering of receptor buds left on the stocks) as those from the 5th node (Figure 4). Clearly, the sensitivity of leaves increases with their ontogenetic rank in *Perilla.* An identical conclusion was reached for the LDP *Lolium* where a given area of an upper leaf is far more effective in induction than the same area of lower leaves.[76] Juvenility was thus ascribed in these two species to the inability of leaves of young plants to respond optimally to inductive cycles. This conclusion would be strengthened if some indication about the relative source (photosynthetic) capacity of the leaves compared in the two species were available.

Despite the limited number of plants investigated in this respect, the situation found in *Perilla* and *Lolium* is sometimes assumed to be common to many plants. It is not of universal occurrence, however, since there are indications that (1) in *Xanthium* and Biloxi soybean, it is the physiological age of the leaf and not its position on the stem which essentially determines its effectiveness[6,377] and (2) in *Anagallis,* leaf efficiency declines with increase in ontogenetic rank.[56]

D. Influence of the Root System

An extensive investigation by Schwabe and Al-Doori of the cause of the juvenile-like condition in cuttings of the woody perennial, *Ribes nigrum,* has disclosed that the factor most directly responsible for the inability to flower is the proximity of lateral buds to the root system.[378] A distance of more than 20 nodes between a bud and the roots is necessary to permit evocation of this bud. The root effect is well demonstrated by the observation that aerial rooting on long adult shoots totally prevents flower initiation. Whether this effect is also entirely or partially responsible for the normal juvenile condition characteristic of other woody perennials is not known with cer-

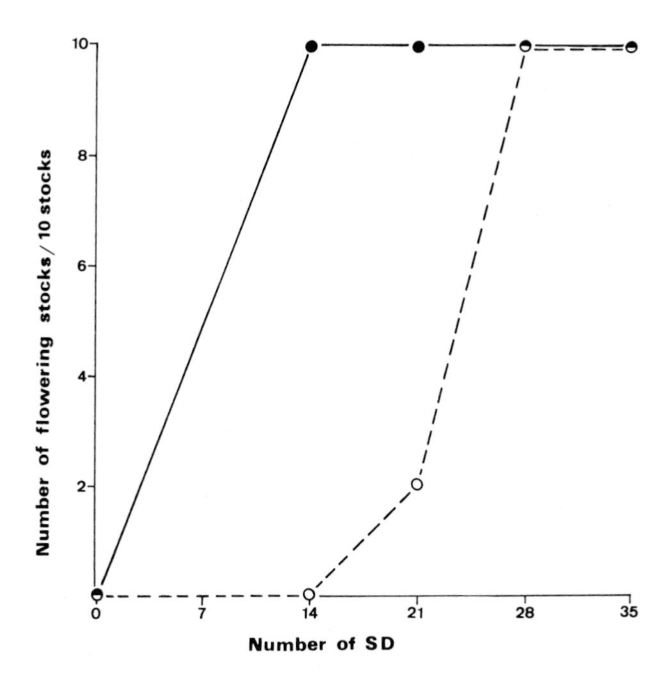

FIGURE 4. Influence of the ontogenetic rank of the donor leaf on the flowering of LD stocks of the SDP *Perilla*. The experimental system is identical to that described in Volume I, Chapter 6, Section II. A. 1. Ten days after grafting of the donor leaf, this leaf is exposed to the indicated number of SD by means of a light-proof bag. ●—● leaf of the 5th pair; O---O leaf of the 2nd pair. (From Zeevaart, J. A. D., *Meded. Landbouwhogesch. Wageningen,* 58(3), 1, 1958. With permission.)

tainty, but an influence of the rootstock on flowering in apples is clearly apparent in Visser's grafting experiments. He showed that the dwarfing rootstock, Malling IX, greatly reduces the length of the juvenile phase in apple seedlings.[379] Another rootstock delays flowering and these effects on precocity are independent of the influence on scion growth rate. It remains to be shown what properties of the shoot system are changed by the different rootstocks.

Root effect in herbaceous plants has not been much studied. In the biennial *Lunaria annua,* removal of the roots shortens the juvenile phase.[104] Derooting also promotes flower formation in several photoperiodic species (see Volume II, Chapter 1); this effect might also come, at least in some cases, e.g., *Scrofularia arguta,* from a shortening of the juvenile phase.

E. Meristem Insensitivity to Floral Promoters

In trees, such as *Citrus* (oranges, grapefruits, etc.) and *Larix* (larch), juvenile scions do not flower more rapidly when grafted onto mature plants bearing flowering shoots,[380,381] suggesting that the meristems of juvenile scions are incapable of responding to stimuli from mature leaves, i.e., they remain juvenile. Robinson and Wareing emphasized that the change-over to the adult phase at the shoot apex in woody plants is thus determined by some mechanism intrinsic to the apex itself, rather than by the conditions prevailing within the differentiated parts of the plants.[381] They postulated that this phase change occurs after the meristematic cells have undergone a certain number of mitotic cycles.

Similar clear-cut evidence in herbaceous species is lacking. A contribution of meris-

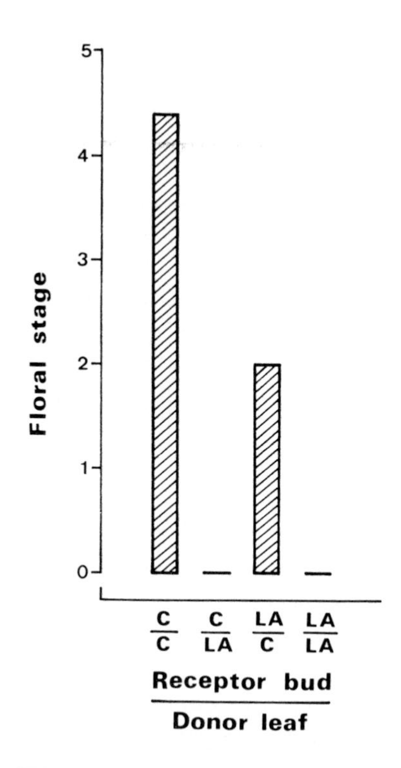

FIGURE 5. Flowering response of plants of *Xanthium strumarium* "synthesized" from two different strains, Chicago (C) and Los Angeles (LA), after exposure to five SD. The flowering response is measured by a score system: 0 = vegetative; 1 = youngest recognizable inflorescence primordium; and 2, 3, 4, and 5 = inflorescence apex with a diameter of 0.25, 0.50, 0.75, and 1 mm, respectively. (Adapted from data of Carpenter, B. H. and Lincoln, R. G., *Science,* 129, 780, 1959.)

tem sensitivity to the overall sensitivity of photoperiodic plants to photoinduction is suggested, however, in reciprocal grafting experiments between early and late strains of these plants. A strain of *Xanthium* collected in the vicinity of Los Angeles is less sensitive to SD cycles than the archetypical Chicago strain of the same species which reacts to a single cycle. Carpenter and Lincoln made reciprocal grafts between the two strains leaving only one leaf on the donor and one bud on the receptor.[382] Induction of the grafted composite with five SD revealed that a leaf of the Chicago strain is far more effective than a leaf of the Los Angeles strain in causing floral initiation in both bud types (Figure 5). In addition, a Los Angeles bud responds less than a Chicago bud to a Chicago leaf. The difference in photoperiodic sensitivity or in earliness of these two strains is thus attributable to differences in *both* the leaf and the bud.

Similar experiments with early and late lines of other photoperiodic species, such as *Pharbitis* and *Kalanchoe* have yielded an identical conclusion;[1,91] so that differences in photoperiodic sensitivity may find their origin not only in the leaves, but also in the meristems. Also, in peas, gene *Lf* which influences the juvenile phase (see above) is believed by Murfet to control the sensitivity of the *apex* to a floral inhibitor and/or a floral promoter, with an *Lf* apex requiring a higher ratio of promoter to inhibitor than an *lf* apex to initiate flowering.[14]

In the biennial *Lunaria annua*, Pierik has observed that, while juvenile plants and shoot cuttings from juvenile plants cannot be vernalized, leaf cuttings from the same plants are fully sensitive to low temperatures (at least if some *de novo* formation of buds occurs prior to vernalization).[104] He concluded that juvenility in *Lunaria* is not a character of the leaves, but is most probably located in the meristems.

CONCLUSIONS

A delay in reproductive development seems necessary if plants are to become large. Thus, juvenility fits immediately as part of an evolutionary strategy concerning plant size and adaptation. The juvenile phase is obviously the result of several physiological systems. The first, found in many annual plants, seems to be related to insufficient photosynthetic capacity and/or another metabolic limitation in the leaves. A second, concerned with the activity of the root system, is present in woody perennials. Whether or not it is also active in herbaceous species remains to be investigated. The third, and potentially most interesting, is that related to the inability of a juvenile meristem to respond to floral stimuli produced in the plant. This kind of juvenility, most readily demonstrated in tree species, but also observed in some herbaceous plants, is what we call *true* juvenility, because it is clearly separable from assimilate supply. It implies that a more or less stable change must occur in meristems as the plants pass from the juvenile to the adult phase. The nature of this change is totally unknown.

ADDENDUM

INTRODUCTION

Information about flower initiation has accumulated since the writing of the manuscript of this volume. We wanted to update our work as much as possible and thus we present in this special section a few selected references which appeared in late 1979 and in 1980.

CHAPTER 2 — CONTROL BY NUTRITION AND WATER STRESS

Flower bud regeneration by thin cellular layers (very small explants composed of epidermal and subepidermal cells) collected from inflorescence branches of day-neutral tobaccos (W38 or Samsun) occurs only when these explants are grown on a solid medium.[382a] Removal of agar is sufficient to prevent flower formation. A similar observation was previously reported with chicory root explant (Section IV of this chapter). Small glass beads, 4 mm in diameter, or a pH increase substitute well for agar. Since in the presence of glass beads, the pH of the culture medium is increased, Cousson and Tran Thanh Van suggest that the promotive effect of the beads is attributable, at least partly, to this change in pH, and has nothing to do with a change in the water potential of the medium.[382a]

CHAPTER 3 — CONTROL BY DAYLENGTH

There is an ever growing number of papers dealing with interactions between daylength and other environmental factors. Several cases were discussed at the Gif conference on flowering.[382b] Confirming an old observation of Resende (see Reference 317), the dual daylength requirement of the LSDP *Bryophyllum daigremontianum* was found to be overcome by growth at the relatively low temperatures of 12 to 17°C. At these temperatures the plant behaves as a typical LDP. It is known since long ago that *Cestrum nocturnum,* another LSDP, will also apparently lose its requirement for SD under certain stress conditions.[382m]

Also, there is a relationship in SD-grown chrysanthemum between the daytime photon flux density and the night break radiant energy required to inhibit flower initiation.[382c] During months of high daytime irradiance (summer), more radiant energy is required for an effective night break than during periods of low daytime irradiance (winter).

High atmospheric CO_2 levels ($\geq 1\%$) perturb completely the photoperiodic behavior of the SDP *Pharbitis.*[382d] Flower initiation is delayed under inductive SD and promoted under noninductive LD. Similar unpublished results were obtained with *Xanthium.*[382d] Hicklenton and Jolliffe conclude from these and other results that, in all cases examined so far, the presence of $\geq 1\%$ CO_2 has acted to oppose normal photoperiodic flowering responses.

CHAPTER 5 — CONTROL BY LOW TEMPERATURE

The effectiveness of a chilling treatment may be adversely affected not only by a subsequent heat treatment as explained in Section V.A. of this chapter, but also by a *previous* heat exposure. Such an effect of high temperatures (25 to 35°C) was recently observed in celery, a biennial species, and might be of economic importance.[382e] A similar action of heat was already known in other cold-requiring species such as winter rye,[250] *Arabidopsis,*[273] and *Cheiranthus,*[240] and called "predevernalization" or "antivernalization".[250,256]

Complex interactions between temperature, daylength, and light intensity were found to influence flower initiation in several cold-requiring species.[382b] In the Shuokan *Chrysanthemum,* for example, the requirement for vernalization can be substituted by growth at high irradiances.

CHAPTER 6 — CLASSICAL THEORIES OF INDUCTION

Various difficulties raised by the florigen theory were stressed in Section II of this chapter. A recent observation of Tanimoto and Harada on the obligate SDP *Perilla* has added aaother complication.[382f] Leaf disks collected from *strictly vegetative* plants were subcultured in noninductive daylength conditions. Discs taken from the upper mature leaves eventually produce *flowering shoots* on a simple medium containing a cytokinin:auxin ratio equal to ten. How can we explain that on the basis of the florigen concept?

New attempts to isolate floral promoters from pea have failed.[382g]

Further evidence for production of a graft-transmissible floral inhibitor by leaves of photoperiodic species held in noninductive daylength conditions was recently given in the cases of the LDP *Nicotiana sylvestris* and the SDP Maryland Mammoth tobacco.[382h,382i] In both species, the inhibitor seems to interact in a simple manner with a floral promoter at the shoot meristem.[382h] Vegetative stem calluses from Trapezond tobacco, i.e., callus regenerating vegetative buds, also release inhibitory material(s) in the culture medium whereas reproductive calluses, i.e., callus regenerating floral buds, produce and release floral promoter(s).[382j] The chemical nature of these water-soluble materials in unknown.

CHAPTER 7 — AGE AND FLOWER INITIATION

Two recent investigations deal with remarkable manipulations of the duration of the juvenile phase. In *Panax ginseng* (ginseng), a species with an extended juvenile phase of 3 years, embryoids formed on cultured root callus have been observed to initiate flowers at the cotyledonary stage.[382k] On the other hand, the apical meristem of the day-neutral W38 tobacco, which normally flowers after producing 30 to 40 nodes, has been maintained in a perpetually vegetative condition simply by rooting the upper portion of the shoot each time the plant has developed six to ten leaves.[382l] Leaf area does not interfere with the regulation of meristem conversion to the reproductive condition in this tobacco, so that it appears that the meristem must be separated from the root system by some minimal number of nodes before the floral transition can occur. This result supports the idea that roots play a role in juvenility not only in some woody perennials, as proposed in Section IV. D. of this chapter, but also in some herbaceous species.

REFERENCES

1. **Imamura, S. I.,** *Physiology of Flowering in Pharbitis nil,* Japanese Society of Plant Physiologists, Tokyo, 1967.
2. **Cumming, B. G.,** *The Induction of Flowering. Some Case Histories,* Evans, L. T., Ed., Macmillan, Melbourne, 1969, 156.
3. **Friend, D. J. C.,** *The Induction of Flowering. Some Case Histories,* Evans, L. T., Ed., Macmillan, Melbourne, 1969, 364.
4. **Kandeler, R.,** *Z. Bot.,* 43, 61, 1955.
5. **Hillman, W. S.,** *The Induction of Flowering. Some Case Histories,* Evans, L. T., Ed., Macmillan, Melbourne, 1969, 186.
6. **Salisbury, F. B.,** *The Flowering Process,* Pergamon Press, Oxford, 1963.
7. **Evans, L. T.,** *The Induction of Flowering. Some Case Histories,* Evans, L. T., Ed., Macmillan, Melbourne, 1969, 328.
8. **Bernier, G.,** *The Induction of Flowering. Some Case Histories,* Evans, L. T., Ed., Macmillan, Melbourne, 1969, 306.
9. **Jacobs, W. P. and Raghavan, V.,** *Phytomorphology,* 12, 144, 1962.
10. **Thomas, R. G.,** *Ann. Bot. (London),* 25, 255, 1961.
11. **Maksymowych, R., Cordero, R. E., and Erickson, R. O.,** *Am. J. Bot.,* 63, 1047, 1976.
12. **Lyndon, R. F.,** *Symp. Soc. Exp. Biol.,* 31, 1977, 221.
13. **Bernier, G., Bodson, M., Havelange, A., Jacqmard, A., and Kinet, J. M.,** unpublished data, 1964-1980.
14. **Murfet, I. C.,** *Annu. Rev. Plant Physiol.,* 28, 253, 1977.
15. **Evans, L. T.,** *The Induction of Flowering. Some Case Histories,* Evans, L. T., Ed., Macmillan, Melbourne, 1969, 457.
16. **Cleland, C. F. and Briggs, W. R.,** *Plant Physiol.,* 42, 1553, 1967.
17. **Salisbury, F. B.,** *Plant Physiol.,* 30, 327, 1955.
18. **Léonard, M., Kinet, J. M., Bodson, M., Havelange, A., Jacqmard, A., and Bernier, G.,** *Plant Physiol.,* in press.
19. **Salisbury, F. B.,** *The Induction of Flowering. Some Case Histories,* Evans, L. T., Ed., Macmillan, Melbourne, 1969, 14.
20. **Evans, L. T.,** *New Phytol.,* 59, 163, 1960.
21. **Lang, A.,** *Encyclopedia of Plant Physiology,* Vol. 15 (Part 1), Ruhland, W., Ed., Springer-Verlag, Berlin, 1965, 1380.
22. **Collins, W. J. and Wilson, J. H.,** *Ann. Bot. (London),* 38, 175, 1974.
23. **Bagnard, C.,** *Can. J. Bot.,* 58, 1138, 1980.
24. Evans, L. T., *The Induction of Flowering. Some Case Histories,* Evans, L. T., Ed., Macmillan, Melbourne, 1969, 1.
25. **Murneek, A. E.,** *Vernalization and Photoperiodism,* Murneek, A. E. and Whyte, R. O., Eds., Chronica Botanica, Waltham, Mass., 1948, 83.
26. **Naylor, A. W.,** *Bot. Gaz. (Chicago),* 103, 342, 1941.
27. **Chailakhyan, M. Kh.,** *C. R. Dokl. Acad. Sci. URSS,* 43, 75, 1944.
28. **El Hinnawy, E. I.,** *Meded. Landbouwhogesch. Wageningen,* 56(9), 1, 1956.
29. **Gott, M. B., Gregory, F. G., and Purvis, O. N.,** *Ann. Bot. (London),* 19, 87, 1955.
30. **Calder, D. M. and Cooper, J. P.,** *Nature (London),* 191, 195, 1961.
31. **Tran Thanh Van-Le Kiem Ngoc, M.,** *Ann. Sci. Nat. Bot. Biol. Veg.,* 12e sér., 6, 373, 1965.
32. **Deltour, R.,** *Cellular and Molecular Aspects of Floral Induction,* Bernier, G., Ed., Longman, London, 1970, 416.
33. **Wada, K.,** *Plant Cell Physiol.,* 15, 381, 1974.
34. **Diomaiuto-Bonnand, J.,** *C. R. Acad. Sci.,* 278, 49, 1974.
35. **Hillman, W. S.,** *Am. J. Bot.,* 48, 413, 1961.
36. **Hillman, W. S.,** *Am. J. Bot.,* 49, 892, 1962.
37. **Pieterse, A. H., Bhalla, P. R., and Sabharwal, P. S.,** *Plant Cell Physiol.,* 11, 879, 1970.
38. **Pieterse, A. H. and Müller, L. J.,** *Plant Cell Physiol.,* 18, 45, 1977.
39. Seth, P. N., Venkataraman, R., and Maheshwari, S. C., *Planta,* 90, 349, 1970.
40. **Graves, C. J. and Sutcliffe, J. F.,** *Ann. Bot. (London),* 38, 729, 1974.
41. **Takimoto, A. and Tanaka, O.,** *Plant Cell Physiol.,* 14, 1133, 1973.
42. **Watanabe, K. and Takimoto, A.,** *Plant Cell Physiol.,* 18, 1369, 1977.
43. **Takimoto, A. and Tanaka, O.,** *Plant Growth Substances 1973,* Hirokawa Publishing, Tokyo, 1974, 953.
44. **Hillman, W. S.,** *Light and Life,* McElroy, W. D. and Glass, B., Eds., Johns Hopkins Press, Baltimore, 1961, 673.

45. **Smith, H. J., McIlrath, W. J., and Bogorad, L.**, *Bot. Gaz. (Chicago)*, 118, 174, 1956.
46. **Gibby, D. D. and Salisbury, F. B.**, *Plant Physiol.*, 47, 784, 1971.
47. **Tanaka, O., Takimoto, A., and Cleland, C. F.**, *Plant Cell Physiol.*, 20, 267, 1979.
48. **Bronchart, R.**, *Mem. Soc. R. Sci. Liege*, 5e sér., 8(2), 1, 1963.
49. **Bouniols, A.**, *Plant Sci. Lett.*, 2, 363, 1974.
50. **Sotta, B.**, *Physiol. Plant.*, 43, 337, 1978.
51. **Aspinall, D. and Husain, I.**, *Aust. J. Biol. Sci.*, 23, 925, 1970.
52. **Murneek, A. E.**, *Vernalization and Photoperiodism*, Murneek, A. E. and Whyte, R. O., Eds., Chronica Botanica, Waltham, Mass., 1948, 39.
53. **Wellensiek, S. J.**, *The Induction of Flowering. Some Case Histories*, Evans, L. T., Ed., Macmillan, Melbourne, 1969, 350.
54. **Hamner, K. C. and Bonner, J.**, *Bot. Gaz. (Chicago)*, 100, 388, 1938.
55. **Vince-Prue, D.**, *Photoperiodism in Plants*, McGraw-Hill, London, 1975.
56. **Ballard, L. A. T.**, *The Induction of Flowering. Some Case Histories*, Evans, L. T., Ed., Macmillan, Melbourne, 1969, 376.
57. **Evans, L. T., Borthwick, H. A., and Hendricks, S. B.**, *Aust. J. Biol. Sci.*, 18, 745, 1965.
58. **Hanke, J., Hartmann, K. M., and Mohr, H.**, *Planta*, 86, 235, 1969.
59. **Takimoto, A.**, *The Induction of Flowering. Some Case Histories*, Evans, L. T., Ed., Macmillan, Melbourne, 1969, 90.
60. **Halaban, R.**, *Plant Physiol.*, 43, 1894, 1968.
61. **Hamner, K. C.**, *Bot. Gaz. (Chicago)*, 101, 658, 1940.
62. **Takimoto, A.**, *Plant Cell Physiol.*, 1, 241, 1960.
63. **Kinet, J. M., Bernier, G., Bodson, M., and Jacqmard, A.**, *Plant Physiol.*, 51, 598, 1973.
64. **Redei, G. P.**, *Genetic Manipulations with Plant Material*, Ledoux, L., Ed., Plenum Press, New York, 1975, 183.
65. **Chailakhyan, M. Kh., Kochankov, V. G., and Muzafarov, B. M.**, *Dokl. Akad. Nauk SSSR*, 234, 252, 1977.
66. **Fife, J. M. and Price, C.**, *Plant Physiol.*, 28, 475, 1953.
67. **Inouye, J., Tashima, Y., and Katayama, T.**, *Plant Cell Physiol.*, 5, 355, 1964.
68. **Knott, J. E.**, *Proc. Am. Soc. Hortic. Sci.*, 31, 152, 1934.
69. **Lona, F.**, *Proc. K. Ned. Akad. Wet.*, 62, 204, 1959.
70. **Zeevaart, J. A. D.**, *Planta*, 98, 190, 1971.
71. **Zeevaart, J. A. D.**, *Meded. Landbouwhogesch. Wageningen*, 58(3), 1, 1958.
72. **Rossini, L. M. E.**, *Cellular and Molecular Aspects of Floral Induction*, Bernier, G., Ed., Longman, London, 1970, 383.
73. **Khudairi, A. K. and Hamner, K. C.**, *Plant Physiol.*, 29, 251, 1954.
74. **Evans, L. T.**, *Aust. J. Biol. Sci.*, 15, 291, 1962.
75. **Jacques, M.**, *Physiol. Veg.*, 9, 461, 1971.
76. **Evans, L. T.**, *Aust. J. Biol. Sci.*, 13, 123, 1960.
77. **Borthwick, H. A. and Parker, M. W.**, *Bot. Gaz. (Chicago)*, 101, 806, 1940.
78. **Friend, D. J. C.**, *Physiol. Plant.*, 21, 990, 1968.
79. **Baldev, B.**, *Ann. Bot. (London)*, 26, 173, 1962.
80. **Butenko, R. G. and Chailakhyan, M. Kh.**, *Dokl. Akad. Nauk SSSR*, 141, 1239, 1961.
81. **Jacobs, W. P. and Suthers, H. B.**, *Am. J. Bot.*, 58, 836, 1971.
82. **Brulfert, J.**, *Rev. Gen. Bot.*, 72, 641, 1965.
83. **Bhargava, S. C.**, *Meded. Landbouwhogesch. Wageningen*, 64(12), 1, 1964.
84. **Nitsch, C. and Nitsch, J. P.**, *Planta*, 72, 371, 1967.
85. **Paulet, P.**, *Rev. Gen. Bot.*, 72, 697, 1965.
86. **Naylor, A. W.**, *Bot. Gaz. (Chicago)*, 102, 557, 1941.
87. **Schwabe, W. W.**, *J. Exp. Bot.*, 10, 317, 1959.
88. **Lincoln, R. G., Raven, K. A., and Hamner, K. C.**, *Bot. Gaz. Chicago*, 119, 179, 1958.
89. **Sachs, R. M.**, *Plant Physiol.*, 31, 185, 1956.
90. **Chailakhyan, M. Kh. and Yanina, L. I.**, *Dokl. Akad. Nauk SSSR*, 199, 234, 1971.
91. **Van de Pol, P. A.**, *Meded. Landbouwhogesch. Wageningen*, 72(9), 1, 1972.
92. **Zeevaart, J. A. D.**, *Annu. Rev. Plant Physiol.*, 27, 321, 1976.
93. **Sironval, C.**, *C. R. Rech. IRSIA*, 18, 1, 1957.
94. **Lam, S. L. and Leopold, A. C.**, *Am. J. Bot.*, 47, 256, 1960.
94a. **Lona, F.**, *Nuovo G. Bot. Ital.*, 53, 548, 1946.
95. **Bonner, J. and Liverman, J.**, *Growth and Differentiation in Plants*, Loomis, W. E., Ed., Iowa State College Press, Ames, 1953, 283.
96. **Zeevaart, J. A. D. and Lang, A.**, *Planta*, 58, 531, 1962.
97. **Wellensiek, S. J.**, *Z. Pflanzenphysiol.*, 55, 1, 1966.

98. **Deronne, M. and Blondon, F.,** *Physiol. Veg.,* 15, 219, 1977.
99. **Zeevaart, J. A. D.,** *The Induction of Flowering. Some Case Histories,* Evans, L. T., Ed., Macmillan, Melbourne, 1969, 116.
100. **Lam, S. L. and Leopold, A. C.,** *Am. J. Bot.,* 48, 306, 1961.
101. **Aghion-Prat, D.,** *Physiol. Veg.,* 3, 229, 1965.
102. **Margara, J.,** *C. R. Acad. Sci.,* 268, 803, 1969.
103. **Nitsch, J. P., Nitsch, C., Rossini, L., Ringe, F., and Harada, H.,** *Cellular and Molecular Aspects of Floral Induction,* Bernier, G., Ed., Longman, London, 1970, 369.
104. **Pierik, R. L. M.,** *Meded. Landbouwhogesch. Wageningen,* 67(6), 1, 1967.
105. **Konstantinova, T. N., Aksenova, N. P., Bavrina, T. V., and Chailakhyan, M.Kh.,** *Dokl. Akad. Nauk SSSR,* 187, 466, 1969.
106. **Braun, A. C.,** *The Biology of Cancer,* Addison-Wesley, Reading, Mass., 1974.
107. **Roberts, R. H. and Struckmeyer, B. E.,** *J. Agric. Res. Washington, D.C.,* 56, 633, 1938.
108. **Went, F.,** *The Experimental Control of Plant Growth,* Chronica Botanica, Waltham, Mass., 1957.
109. **Nitsch, J. P. and Went, F. W.,** *Photoperiodism and Related Phenomena in Plants and Animals,* Withrow, R. B., Ed., American Association for the Advancement of Science, Washington, D.C., 1959, 311.
110. **Ahmed, G. E. D. F. and Jacques, M.,** *C.R. Acad. Sci.,* 280, 617, 1975.
111. **Shinozaki, M.,** *Abstr. 10th Int. Conf. on Plant Growth Substances,* Madison, Wis., 1979, 50.
112. **Chouard, P.,** *Annu. Rev. Plant Physiol.,* 11, 191, 1960.
113. **Wellensiek, S. J.,** *Meded. Landbouwhogesch. Wageningen,* 60(7), 1, 1960.
114. **Ketellapper, H. J. and Barbaro, A.,** *Phyton (Buenos Aires),* 23, 33, 1966.
115. **Murneek, A. E.,** *Bot. Gaz. (Chicago),* 102, 269, 1940.
116. **Wellensiek, S. J.,** *Acta Bot. Neerl.,* 17, 5, 1968.
117. **de Zeeuw, D.,** *Proc. K. Ned. Akad. Wet.,* 56, 418, 1953.
118. **Meijer, G.,** *Acta Bot. Neerl.,* 8, 189, 1959.
119. **Takimoto, A.,** *Plant Cell Physiol.,* 14, 1217, 1973.
120. **Kandeler, R., Hügel, B., and Rottenburg, Th.,** *Environmental and Biological Control of Photosynthesis,* Marcelle, R., Ed., Dr W. Junk, The Hague, Netherlands, 1975, 161.
121. **Bodson, M., King, R. W., Evans, L. T., and Bernier, G.,** *Aust. J. Plant Physiol.,* 4, 467, 1977.
122. **Friend, D. J. C., Deputy, J., and Quedado, R.,** *Photosynthesis and Plant Development,* Marcelle, R., Clijsters, H., and Van Poucke, M., Eds., Dr. W. Junk, The Hague, Netherlands, 1979, 59.
123. **Sachs, R. M.,** *Colloq. Int. C.N.R.S.,* 285, 169, 1979.
124. **Bavrina, T. V., Aksenova, N. P., and Konstantinova, T. N.,** *Fiziol. Rast.,* 16, 381, 1969.
125. **Quedado, R. M. and Friend, D. J. C.,** *Plant Physiol.,* 62, 802, 1978.
126. **Posner, H.,** *Plant Physiol.,* 48, 361, 1971.
127. **Purohit, A. N. and Tregunna, E. B.,** *Can. J. Bot.,* 52, 1283, 1974.
128. **Moshkov, B. S. and Odumanova-Dunaeva, G. A.,** *Dokl. Akad. Nauk SSSR,* 203, 714, 1972.
129. **Bassi, P. K., Tregunna, E. B., and Purohit, A. N.,** *Plant Physiol.,* 56, 335, 1975.
130. **Evans, L. T.,** *Aust. J. Biol. Sci.,* 15, 281, 1962.
131. **Chailakhyan, M. Kh. and Konstantinova, T. N.,** *Fiziol. Rast.,* 9, 693, 1962.
132. **Chouard, P.,** *Mem. Soc. Bot. Fr.,* 96, 1949.
133. **Parker, M. W., Hendricks, S. B., Borthwick, H. A., and Scully, N. J.,** *Bot. Gaz. (Chicago),* 108, 1, 1946.
134. **Borthwick, H. A., Hendricks, S. B., and Parker, M. W.,** *Bot. Gaz. (Chicago),* 110, 103, 1948.
135. **Parker, M. W., Hendricks, S. B., and Borthwick, H. A.,** *Bot. Gaz. (Chicago),* 111, 242, 1950.
136. **Borthwick, H. A., Hendricks, S. B., and Parker, M. W.,** *Proc. Natl. Acad. Sci. U.S.A.,* 38, 929, 1952.
137. **Downs, R. J.,** *Plant Physiol.,* 31, 279, 1956.
138. **Butler, W. L., Norris, K. H., Siegelman, H. W., and Hendricks, S. B.,** *Proc. Natl. Acad. Sci. U.S.A.,* 45, 1703, 1959.
139. **Pratt, L. H.,** *Photochem. Photobiol.,* 27, 81, 1978.
140. **Kendrick, R. E. and Spruit, C. J. P.,** *Photochem. Photobiol.,* 26, 201, 1977.
141. **Mumford, F. E. and Jenner, E. L.,** *Biochemistry,* 5, 3657, 1966.
142. **Cathey, H. M. and Borthwick, H. A.,** *Bot. Gaz. (Chicago),* 119, 71, 1957.
143. **Fredericq, H.,** *Plant Physiol.,* 39, 812, 1964.
144. **Kasperbauer, M. J., Borthwick, H. A., and Hendricks, S. B.,** *Bot. Gaz. (Chicago),* 124, 444, 1963.
145. **Borthwick, H. A. and Downs, R. J.,** *Bot. Gaz. (Chicago),* 125, 227, 1964.
146. **Borthwick, H. A.,** *Am. Nat.,* 98, 347, 1964.
147. **Fredericq, H.,** *Biol. Jaarb.,* 33, 66, 1965.
148. **Nakayama, S., Borthwick, H. A., and Hendricks, S. B.,** *Bot. Gaz. (Chicago),* 121, 237, 1960.
149. **Mancinelli, A. L. and Downs, R. J.,** *Plant Physiol.,* 42, 95, 1967.

150. **Cumming, B. G.,** *Can. J. Bot.,* 41, 901, 1963.
151. **Purves, W. K.,** *Planta,* 56, 684, 1961.
152. **Cathey, H. M.,** *The Induction of Flowering. Some Case Histories,* Evans L. T., Ed., Macmillan, Melbourne, 1969, 268.
153. **Evans, L. T. and King, R. W.,** *Z. Pflanzenphysiol.,* 60, 277, 1969.
154. **Cumming, B. G., Hendricks, S. B., and Borthwick, H. A.,** *Can. J. Bot.,* 43, 825, 1965.
155. **King, R. W. and Cumming, B. G.,** *Planta,* 108, 39, 1972.
156. **King, R. W., Vince-Prue, D., and Quail, P. H.,** *Planta,* 141, 15, 1978.
157. **Salisbury, F. B.,** *Planta,* 66, 1, 1965.
158. **King, R. W.,** *Aust. J. Plant Physiol.,* 1, 445, 1974.
159. **Marushige, K. and Marushige, Y.,** *Bot. Mag.,* 79, 397, 1966.
160. **Friend, D. J. C.,** *Physiol. Plant.,* 35, 286, 1975.
161. **Lane, H. C., Cathey, H. M., and Evans, L. T.,** *Am. J. Bot.,* 52, 1006, 1965.
162. **Harris, G. P.,** *Ann. Bot. (London),* 32, 187, 1968.
163. **Stolwijk, J. A. J. and Zeevaart, J. A. D.,** *Proc. K. Ned. Akad. Wet.,* 58, 386, 1955.
164. **Vince, D.,** *Physiol. Plant.,* 18, 474, 1965.
165. **Ishiguri, Y. and Oda, Y.,** *Plant Cell Physiol.,* 13, 131, 1972.
166. **Meijer, G. and Van der Veen, R.,** *Acta Bot. Neerl.,* 6, 429, 1957.
167. **Schneider, M. J., Borthwick, H. A., and Hendricks, S. B.,** *Am. J. Bot.,* 54, 1241, 1967.
168. **Borthwick, H. A., Hendricks, S. B., Schneider, M. J., Taylorson, R. B., and Toole, V. K.,** *Proc. Natl. Acad. Sci. U.S.A.,* 64, 479, 1969.
169. **Takimoto, A.,** *Bot. Mag.,* 70, 312, 1957.
170. **Friend, D. J. C.,** *Physiol. Plant.,* 17, 909, 1964.
171. **Esashi, Y. and Oda, Y.,** *Plant Cell Physiol.,* 7, 59, 1966.
172. **Deitzer, G. F., Hayes, R., and Jabben, M.,** *Plant Physiol.,* 64, 1015, 1979.
173. **Friend, D. J. C.,** *Physiol. Plant.,* 21, 1185, 1968.
174. **Jacques, M. and Jacques, R.,** *C.R. Acad. Sci.,* 269, 2107, 1969.
175. **Blondon, F. and Jacques, R.,** *C.R. Acad. Sci.,* 270, 947, 1970.
176. **Imhoff, C., Brulfert, J., and Jacques, R.,** *C.R. Acad. Sci.* 273, 737, 1971.
177. **Holland, R. W. K. and Vince, D.,** *Nature (London),* 219, 511, 1968.
178. **Tcha, K. H., Jacques, R., and Jacques, M.,** *C.R. Acad. Sci.,* 283, 341, 1976.
179. **Funke, G. L.,** *Vernalization and Photoperiodism,* Murneek, A. E. and Whyte, R. O., Eds., Chronica Botanica, Waltham, Mass., 1948, 79.
180. **Wassink, E. C., Sluijsmans, C. M. J., and Stolwijk, J. A. J.,** *Proc. K. Ned. Akad. Wet.,* 53, 1466, 1950.
181. **Jacques, M. and Jacques, R.,** *C.R. Acad. Sci.,* 287, 1333, 1978.
182. **Imhoff, C., Lecharny, A., Jacques, R., and Brulfert, J.,** *Plant, Cell Environ.,* 2, 67, 1979.
183. **Evans, L. T.,** *Aust. J. Plant Physiol.,* 3, 207, 1976.
184. **Holland, R. W. K. and Vince, D.,** *Planta,* 98, 232, 1971.
185. **Ishiguri, Y. and Oda, Y.,** *Plant Cell Physiol.,* 15, 287, 1974.
186. **Shropshire, W. A.,** *Photophysiology,* Vol. 7, Giese, A. C., Ed., Academic Press, New York, 1972, 34.
187. **Johnson, C. B. and Tasker, R.,** *Plant, Cell Environ.,* 2, 259, 1979.
188. **Jose, A. M. and Vince-Prue, D.,** *Photochem. Photobiol.,* 27, 209, 1978.
189. **Pittendrigh, C. S.,** *Z. Pflanzenphysiol.,* 54, 275, 1966.
190. **Takimoto, A. and Hamner, K. C.,** *Plant Physiol.,* 39, 1024, 1964.
191. **King, R. W. and Cumming, B. G.,** *Planta,* 103, 281, 1972.
192. **Salisbury, F. B.,** *Planta,* 59, 518, 1963.
193. **King, R. W.,** *Aust. J. Plant Physiol.,* 6, 417, 1979.
194. **Bünning, E.,** *Ber. Dtsch. Bot. Ges.,* 54, 590, 1936.
195. **Bünning, E.,** *Cold Spring Harbor Symp. Quant. Biol.,* 25, 249, 1960.
196. **Bünning, E.,** *Photochem. Photobiol.,* 9, 219, 1969.
197. **Pittendrigh, C. S.,** *Proc. Natl. Acad. Sci. U.S.A.,* 69, 2734, 1972.
198. **Blaney, L. T. and Hamner, K. C.,** *Bot. Gaz. (Chicago)* 119, 10, 1957.
199. **Nanda, K. K. and Hamner, K. C.,** *Bot. Gaz. (Chicago),* 120, 14, 1958.
200. **Shibata, O. and Takimoto, A.,** *Plant Cell Physiol.,* 16, 513, 1975.
201. **Hillman, W. S.,** *Am. Nat.,* 98, 323, 1964.
202. **Engelmann, W.,** *Planta,* 55, 496, 1960.
203. **Hsu, J. C. S. and Hamner, K. C.,** *Plant Physiol.,* 42, 725, 1967.
204. **Moore, P. H., Reid, H. B., and Hamner, K. C.,** *Plant Physiol.,* 42, 503, 1967.
205. **Bünsow, R. C.,** *Cold Spring Harbor Symp. Quant. Biol.,* 25, 257, 1960.
206. **Wareing, P. F.,** *Physiol. Plant.,* 7, 157, 1954.
207. **Carpenter, B. H. and Hamner, K. C.,** *Plant Physiol.,* 38, 698, 1963.

208. **Carr. D. J.,** *Physiol. Plant.*, 5, 70, 1952.
209. **Claes, H. and Lang, A.,** *Z. Naturforsch.*, 2b, 56, 1947.
210. **Kinet, J. M.,** *Nature (London)*, 236, 406, 1972.
211. **Coulter, M. W. and Hamner, K. C.,** *Plant Physiol.*, 39, 848, 1964.
212. **Melchers, G.,** *Z. Naturforsch.*, 11b, 544, 1956.
213. **Hussey, G.,** *Physiol. Plant.*, 7, 253, 1954.
214. **Oda, Y.,** *Plant Cell Physiol.*, 10, 399, 1969.
215. **Oota, Y.,** *Plant Cell Physiol.*, 16, 885, 1975.
216. **Oota, Y. and Tsudzuki, T.,** *Plant Cell Physiol.*, 20, 725, 1979.
217. **Pittendrigh, C. S. and Minis, D. H.,** *Am. Nat.*, 98, 261, 1964.
218. **Spector, C. and Paraska, J. R.,** *Physiol. Plant.*, 29, 402, 1973.
219. **Paraska, J. R. and Spector, C.,** *Physiol. Plant.*, 32, 62, 1974.
220. **Takimoto, A. and Hamner, K. C.,** *Plant Physiol.*, 40, 855, 1965.
221. **Borthwick, H. A. and Hendricks, S. B.,** *Science*, 132, 1223, 1960.
222. **Hendricks, S. B.,** *Cold Spring Harbor Symp. Quant. Biol.*, 25, 245, 1960.
223. **Shumate, W. H., Reid, H. B., and Hamner, K. C.,** *Plant Physiol.*, 42, 1511, 1967.
224. **Takimoto, A. and Hamner, K. C.,** *Plant Physiol.*, 40, 859, 1965.
225. **Kato, A.,** *Plant Cell Physiol.*, 20, 1273, 1979.
226. **Kato, A.,** *Plant Cell Physiol.*, 20, 1285, 1979.
227. **Vince, D.,** *Colloq. Int. C.N.R.S.*, 285, 91, 1979.
228. **Hillman, W. S.,** *Light and Plant Development*, Smith, H., Ed., Butterworths, London, 1976, 383.
229. **Hillman, W. S.,** *Proc. Natl. Acad. Sci. U.S.A.*, 73, 501, 1976.
230. **Salisbury, F. B. and Denney, A.,** *Biochronometry*, Menaker, M., Ed., National Academy of Science, Washington D.C., 1971, 292.
231. **Brest, D. E., Hoshizaki, T., and Hamner, K. C.,** *Plant Physiol.*, 47, 676, 1971.
232. **Bollig, I.,** *Z. Pflanzenphysiol.*, 77, 54, 1975.
233. **Bollig, I.,** *Planta*, 135, 137, 1977.
234. **Wagner, E. and Cumming, B. G.,** *Can. J. Bot.*, 48, 1, 1970.
235. **Gressel, J., Zilberstein, A., Porath, D., and Arzee, T.,** *Photoreceptors and Plant Development*, De Greef, J., Ed., Antwerpen University Press, Belgium, 1980, 525.
236. **King, R. W.,** *Can. J. Bot.*, 53, 2631, 1975.
237. **Vince-Prue, D.,** *Light and Plant Development*, Smith, H., Ed., Butterworths, London, 1976, 347.
238. **Whyte, R. O.,** *Vernalization and Photoperiodism*, Murneek, A. E. and Whyte, R. O., Eds., Chronica Botanica, Waltham, Mass., 1948, 1.
239. **Hartsema, A. M.,** *Encyclopedia of Plant Physiology*, Vol. 16, Ruhland, W., Ed., Springer-Verlag, Berlin, 1961, 123.
240. **Barendse, G. W. M.,** *Meded. Landbouwhogesch. Wageningen*, 64(14), 1, 1964.
241. **Picard, C.,** *Ann. Sci. Nat. Bot. Biol. Veg.*, 12e sér., 6, 197, 1965.
242. **Hiller, L. K. and Kelly, W. C.,** *J. Am. Soc. Hortic. Sci.*, 104, 253, 1979.
243. **Wycherley, P. R.,** *Meded. Landbouwhogesch. Wageningen*, 52(2), 75, 1952.
244. **Heide, O. M.,** *Meld. Norg. Landbrukshoegsk.*, 59, 1, 1980.
245. **Sarkar, S.,** *Biol. Zentralbl.*, 77, 1, 1958.
246. **Emsweller, S. L. and Borthwick, H. A.,** *Proc. Am. Soc. Hortic. Sci.*, 35, 755, 1937.
247. **Thompson, H. C.,** *Proc. Am. Soc. Hortic. Sci.*, 45, 425, 1944.
248. **Parlevliet, J. E.,** *Z. Pflanzenphysiol.*, 58, 76, 1967.
249. **Purvis, O. N.,** *Ann. Bot. (London)*, 12, 183, 1948.
250. **Purvis, O. N.,** *Encyclopedia of Plant Physiology*, Vol. 16, Ruhland, W., Ed., Springer-Verlag, Berlin, 1961, 76.
251. **Evans, L. T.,** *J. Agric. Sci.*, 54, 410, 1960.
252. **Hackett, W. P. and Hartmann, H. T.,** *Physiol. Plant.*, 20, 430, 1967.
253. **Schwabe, W. W.,** *J. Exp. Bot.*, 1, 329, 1950.
254. **Purvis, O. N. and Gregory, F. G.,** *Ann. Bot. (London)*, 16, 1, 1952.
255. **Heide, O. M.,** *Meld. Norg. Landbrukshoegsk.*, 49, 1, 1970.
256. **Napp-Zinn, K.,** *Temperature and Life*, Precht, H., Christophersen, J., Hensel, H., and Larcher, W., Eds., Springer-Verlag, Berlin, 1973, 171.
257. **Curtis, O. F. and Chang, H. T.,** *Am. J. Bot.*, Suppl. 17, 1047, 1930.
258. **Chroboczek, E.,** *Mem. Cornell Univ. Agric. Exp. Stn.*, 154, 1, 1934.
259. **Schwabe, W. W.,** *J. Exp. Bot.*, 5, 389, 1954.
260. **Hackett, W. P. and Hartmann, H. T.,** *Bot. Gaz. (Chicago)*, 125, 65, 1964.
261. **Harada, H.,** *Rev. Gen. Bot.*, 69, 201, 1962.
262. **Kruzhilin, A. S. and Shvedskaya, Z. M.,** *Dokl. Akad. Nauk SSSR*, 121, 561, 1958.
263. **Purvis, O. N.,** *Nature (London)*, 145, 462, 1940.

264. **Purvis, O. N.,** *Ann. Bot. (London),* 8, 285, 1944.
265. **Kruzhilin, A. S. and Shvedskaya, Z. M.,** *Fiziol. Rast.,* 7, 287, 1960.
266. **Wellensiek, S. J.,** *Plant Physiol.,* 39, 832, 1964.
267. **Pierik, R. L. M.,** *Naturwissenschaften,* 53, 387, 1966.
268. **Margara, J.,** *C. R. Acad. Sci.,* 278, 2283, 1974.
269. **Grif, V. G.,** *Fiziol. Rast.,* 5, 524, 1958.
270. **Kimball, S. L. and Salisbury, F. B.,** *Bot. Gaz. (Chicago),* 135, 147, 1974.
270a. **Gregory, F. G. and Purvis, O. N.,** *Nature (London),* 138, 1013, 1936.
271. **Waterschoot, H. F.,** *Proc. K. Ned. Akad. Wet.,* 60, 318, 1957.
272. **Schwabe, W. W.,** *J. Exp. Bot.,* 8, 220, 1957.
273. **Napp-Zinn, K.,** *The Induction of Flowering. Some Case Histories,* Evans, L. T., Ed., Macmillan, Melbourne, 1969, 291.
274. **Marks, M. K. and Prince, S. D.,** *New Phytol.,* 82, 357, 1979.
275. **Reid, J. B. and Murfet, I. C.,** *Ann. Bot. (London),* 42, 945, 1977.
276. **Pierik, R. L. M.,** *Z. Pflanzenphysiol.,* 56, 141, 1967.
277. **Chouard, P. and Larrieu, C.,** *C. R. Acad. Sci.,* 259, 2121, 1964.
278. **Blondon, F.,** *C. R. Acad. Sci.,* 272, 2896, 1971.
279. **Mathon, C. C.,** *Phyton (Buenos Aires),* 14, 167, 1960.
280. **Peterson, M. L. and Loomis, W. E.,** *Plant Physiol.,* 24, 31, 1949.
281. **Newell, L. C.,** *Agron. J.,* 43, 417, 1951.
282. **Gardner, F. P. and Loomis, W. E.,** *Plant Physiol.,* 28, 201, 1953.
283. **Blondon, F.,** *Phytotronique et Prospective Horticole. Phytotronique II,* Chouard, P. and de Bilderling, N., Eds., Gauthier-Villars, Paris, 1972, 135.
284. **Margara, J.,** *Ann. Amelior. Plant.,* 10, 361, 1960.
285. **Kasperbauer, M. J., Gardner, F. P., and Loomis, W. E.,** *Plant Physiol.,* 37, 165, 1962.
286. **Caso, O. H. and Kefford, N. P.,** *Aust. J. Biol. Sci.,* 21, 883, 1968.
287. **Gregory, F. G. and Purvis, O. N.,** *Ann. Bot. (London),* 2, 753, 1938.
288. **Napp-Zinn, K.,** *Naturwissenschaften,* 49, 473, 1962.
289. **Haupt, W. and Nakamura, E.,** *Z. Pflanzenphysiol.,* 62, 270, 1970.
290. **Wellensiek, S. J.,** *Proc. K. Ned. Akad. Wet.,* 61, 561, 1958.
291. **Evans, L. T.,** *Grasses and Grasslands,* Barnard, C., Ed., Mcmillan, London, 1964, 126.
292. **Ofir, M. and Koller, D.,** *Aust. J. Plant Physiol.,* 1, 259, 1974.
293. **Sen, B. and Chakravarti, S. C.,** *Nature (London),* 157, 266, 1946.
294. **Gregory, F. G. and Purvis, O. N.,** *Ann. Bot. (London),* 2, 237, 1938.
295. **Reid, J. B.,** *Ann. Bot. (London),* 44, 461, 1979.
296. **Lang, A.,** *Annu. Rev. Plant Physiol.,* 3, 265, 1952.
297. **Chailakhyan, M. Kh.,** *C. R. Dokl. Acad. Sci. URSS,* 16, 227, 1937.
298. **Chailakhyan, M. Kh., Khazhakyan, Kh. K., and Agamyan, L. B.,** *Dokl. Akad. Nauk SSSR,* 230, 1002, 1976.
299. **Zeevaart, J. A. D.,** *Planta,* 140, 289, 1978.
300. **De Stigter, H. C. M.,** *Z. Pflanzenphysiol.,* 55, 11, 1966.
301. **Melchers, G.,** *Ber. Dtsch. Bot. Ges.,* 57, 29, 1939.
302. **Hillman, W. S.,** *The Physiology of Flowering,* Holt, Rinehart and Winston, New York, 1964.
303. **Jacques, M.,** *C. R. Acad. Sci.,* 276, 1705, 1973.
304. **Griesel, W. O.,** *Plant Physiol.,* 38, 479, 1963.
305. **Wellensiek, S. J.,** *Z. Pflanzenphysiol.,* 63, 25, 1970.
306. **Fratianne, D. G.,** *Am. J. Bot.,* 52, 556, 1965.
307. **Zeevaart, J. A. D.,** *Plant Physiol.,* 37, 296, 1962.
308. **Bernier, G., Bodson, M., Kinet, J. M., Jacqmard, A., and Havelange, A.,** *Plant Growth Substances 1973,* Hirokawa Publishing, Tokyo, 1974, 980.
309. **Evans, L. T. and Wardlaw, I. F.,** *Planta,* 68, 310, 1966.
310. **Takeba, G. and Takimoto, A.,** *Bot. Mag.,* 79, 811, 1966.
311. **King, R. W., Evans, L. T., and Wardlaw, I. F.,** *Z. Pflanzenphysiol.,* 59, 377, 1968.
312. **Kavon, D. L. and Zeevaart, J. A. D.,** *Planta,* 144, 201, 1979.
313. **Galston, A. W.,** *Bot. Gaz. (Chicago),* 110, 495, 1949.
314. **Sironval, C.,** *Mem. Acad. R. Belg., Cl. Sci.,* 26(4), 1, 1951.
315. **Lincoln, R. G., Raven, K. A., and Hamner, K. C.,** *Bot. Gaz. (Chicago),* 117, 193, 1956.
316. **Harder, R.,** *Symp. Soc. Exp. Biol.,* 2, 117, 1948.
317. **Zeevaart, J. A. D.,** *The Induction of Flowering. Some Case Histories,* Evans, L. T., Ed., Macmillan, Melbourne, 1969, 435.
318. **King, R. W. and Zeevaart, J. A. D.,** *Plant Physiol.,* 51, 727, 1973.
319. **Chailakhyan, M. Kh. and Butenko, R. G.,** *Fiziol. Rast.,* 4, 450, 1957.

320. **Zeevaart, J. A. D., Brede, J. M., and Cetas, C. B.,** *Plant Physiol.*, 60, 747, 1977.
321. **Biswas, P. K., Paul, K. B., and Henderson, J. H. M.,** *Physiol. Plant.*, 19, 875, 1966.
322. **Tomita, T.,** *Biochemistry and Physiology of Plant Growth Substances*, Wightman, F. and Setterfield, G., Eds., Runge Press, Ottawa, 1968, 1399.
323. **Chailakhyan, M. Kh., Grigorieva, N. Y., and Lozhnikova, V. N.,** *Dokl. Akad. Nauk SSSR*, 236, 773, 1977.
324. **Lincoln, R. G., Mayfield, D. L., and Cunningham, A.,** *Science*, 133, 756, 1960.
325. **Mayfield, D. L.,** *Colloq. Int. C.N.R.S.*, 123, 621, 1964.
326. **Lincoln, R. G., Cunningham, A., and Hamner, K. C.,** *Nature (London)*, 202, 559, 1964.
327. **Carr, D. J.,** *Ann. N.Y. Acad. Sci.*, 144, 305, 1967.
328. **Hodson, H. K. and Hamner, K. C.,** *Science*, 167, 384, 1970.
329. **Cleland, C. F.,** *Plant Physiol.*, 54, 899, 1974.
330. **Cleland, C. F. and Ajami, A.,** *Plant Physiol.*, 54, 904, 1974.
331. **Raghavan, V. and Jacobs, W. P.,** *Am. J. Bot.*, 48, 751, 1961.
332. **Lang, A.,** *Naturwissenschaften*, 43, 284, 1956.
333. **Chailakhyan, M. Kh.,** *Biol. Zentralbl.*, 77, 641, 1958.
334. **Lona, F.,** *Nuovo G. Bot. Ital.*, 56, 479, 1949.
335. **von Denffer,,D.,** *Naturwissenschaften*, 37, 296, 317, 1950.
336. **Withrow, A. P., Withrow, R. B., and Biebel, J. P.,** *Plant Physiol.*, 18, 294, 1943.
337. **Paton, D. M. and Barber, H. N.,** *Aust. J. Biol. Sci.*, 8, 230, 1955.
338. **Haupt, W.,** *The Induction of Flowering. Some Case Histories*, Evans, L. T., Ed., Macmillan, Melbourne, 1969, 393.
339. **Reid, J. B. and Murfet, I. C.,** *J. Exp. Bot.*, 28, 811, 1977.
340. **Murfet, I. C. and Reid, J. B.,** *Aust. J. Biol. Sci.*, 26, 675, 1973.
341. **Lang, A., Chailakhyan, M. Kh., and Frolova, I. A.,** *Proc. Natl. Acad. Sci. U.S.A.*, 74, 2412, 1977.
342. **Jacobs, W. P.,** *Plant Growth Substances 1979*, Skoog, F., Ed., Springer-Verlag, Berlin, 1980, 301.
343. **Guttridge, C. G.,** *Nature (London)*, 178, 50, 1956.
344. **Guttridge, C. G.,** *Ann. Bot. (London)*, 23, 351, 1959.
345. **Thompson, P. A. and Guttridge, C. G.,** *Ann. Bot. (London)*, 24, 482, 1960.
346. **Evans, L. T.,** *Aust. J. Biol. Sci.*, 13, 429, 1960.
347. **Evans, L. T. and Wardlaw, I. F.,** *Aust. J. Biol. Sci.*, 17, 1, 1964.
348. **Searle, N. E.,** *Plant Physiol.*, 40, 261, 1965.
349. **Lang, A.,** *Cellular and Molecular Aspects of Floral Induction*, Bernier, G., Ed., Longman, London, 1970, 302.
350. **Schwabe, W. W.,** *Planta*, 103, 18, 1972.
351. **Blake, J.,** *Planta*, 103, 126, 1972.
352. **Wareing, P. F. and El-Antably, H. M. M.,** *Cellular and Molecular Aspects of Floral Induction*, Bernier, G., Ed., Longman, London, 1970, 285.
353. **Bernier, G.,** *Mem. Acad. R. Belg., Clin. Sci.*, 16(1), 1, 1964.
354. **Hillman, W. S.,** *Am. J. Bot.*, 62, 537, 1975.
355. **Gott, M. B.,** *Nature (London)*, 180, 714, 1957.
356. **Napp-Zinn, K.,** *Z. Bot.*, 45, 379, 1957.
357. **Prince, S. D., Marks, M. K., and Carter, R. N.,** *New Phytol.*, 81, 265, 1978.
358. **Schwabe, W. W.,** *The Induction of Flowering. Some Case Histories*, Evans, L. T., Ed., Macmillan, Melbourne, 1969, 227.
359. **Holdsworth, M.,** *J. Exp. Bot.*, 7, 395, 1956.
360. **Galinat, W. C. and Naylor, A. W.,** *Am. J. Bot.*, 38, 38, 1951.
361. **Van Rossem, A. and Bolhuis, G. G.,** *Neth. J. Agric. Sci.*, 2, 302, 1954.
362. **Wellensiek, S. J.,** *Z. Pflanzenphysiol.*, 61, 462, 1969.
363. **Cockshull, K. E.,** *J. Hortic. Sci.*, 51, 441, 1976.
364. **Wellensiek, S. J.,** *Etudes de Biologie Végétale. Hommage au Professeur Pierre Chouard*, Jacques, R., Ed., Paris, 1976, 301.
365. **Vergara, B. S., Chang, T. T., and Lilis, R.,** *Int. Rice Res. Inst. Tech. Bull.*, 8, 1, 1969.
366. **Visser, T. and De Vries, D. P.,** *Euphytica*, 19, 141, 1970.
367. **Higazy, M. K. M. T.,** *Meded. Landbouwhogesch. Wageningen*, 62(8), 1, 1962.
368. **Pawar, S. S. and Thompson, H. C.,** *Proc. Am. Soc. Hortic. Sci.*, 55, 367, 1950.
369. **Fisher, J. E.,** *Bot. Gaz. (Chicago)*, 117, 156, 1955.
370. **Chouard, P. and Lourtioux, A.,** *C. R. Acad. Sci.*, 249, 889, 1959.
371. **Hussey, G.,** *J. Exp. Bot.*, 14, 326, 1963.
372. **Jennings, P. R. and Zuck, R. K.,** *Bot. Gaz. (Chicago)*, 116, 199, 1954.
373. **Chailakhyan, M. Kh. and Podol'nyi, V. Z.,** *Fiziol. Rast.*, 15, 949, 1968.
374. **Kujirai, C. and Imamura, S. I.,** *Bot. Mag.*, 71, 408, 1958.

375. **Zeevaart, J. A. D.,** *Planta,* 58, 543, 1962.
376. **Penner, J.,** *Planta,* 55, 542, 1960.
377. **Hamner, K. C.,** *The Induction of Flowering. Some Case Histories,* Evans, L. T., Ed., Macmillan, Melbourne, 1969, 62.
378. **Schwabe, W. W. and Al-Doory, A. H.,** *J. Exp. Bot.,* 24, 969, 1973.
379. **Visser, T.,** *J. Am. Soc. Hortic. Sci.,* 98, 26, 1973.
380. **Furr, J. R., Cooper, W. C., and Reece, P. C.,** *Am. J. Bot.,* 34, 1, 1947.
381. **Robinson, L. W. and Wareing, P. F.,** *New Phytol.,* 68, 67, 1969.
382. **Carpenter, B. H. and Lincoln, R. G.,** *Science,* 129, 780, 1959.
382a. **Cousson, A. and Tran Thanh Van, K.,** *Physiol. Plant.,* 51, 77, 1981.
382b. **Zeevaart, J. A. D.,** *Colloq. Int. C.N.R.S.,* 285, 59, 1979.
382c. **Sachs, R. M., Kofranek, A. M., and Kubota, J.,** *HortScience,* 15, 609, 1980.
382d. **Hicklenton, P. R. and Jolliffe, P. A.,** *Plant Physiol.,* 66, 13, 1980.
382e. **Sachs, M. and Rylski, I.,** *Sci. Hort.,* 12, 231, 1980.
382f. **Tanimoto, S. and Harada, H.,** *Ann. Bot. (London),* 45, 321, 1980.
382g. **Reid, J. B. and Murfet, I. C.,** *Ann. Bot. (London),* 45, 583, 1980.
382h. **Lang, A.,** In *Plant Growth Substances 1979,* Skoog, F., Ed., Springer-Verlag, Berlin, 1980, 310.
382i. **Chailakhyan, M. Kh., Yanina, L. I., and Lotova, G. N.,** *Dokl. Nauk. Akad. SSSR,* 248, 1513, 1979.
382j. **Aksenova, N. P., Konstantinova, T. N., and Bavrina, T. V.,** *Fiziol. Rast.,* 26, 1215, 1979.
382k. **Chang, W.-C. and Hsing, Y.-I.,** *Nature (London),* 284, 341, 1980.
382l. **McDaniel, C. N.,** *Planta,* 148, 462, 1980.
382m. **Sachs, R. M.,** unpublished observations, 1969.
383. **Salisbury, F. B. and Bonner, J.,** *Plant Physiol.,* 31, 141, 1956.
384. **Vince, D. and Mason, D. T.,** *Nature (London),* 174, 842, 1954.
385. **Goh, C. J. and Seetoh, H. C.,** *Ann. Bot. (London),* 37, 113, 1973.
386. **Goh, C. J.,** *Ann. Bot. (London),* 39, 931, 1975.

GLOSSARY

Ambiphotoperiodic plants	Plants in which flower initiation is promoted by relatively short or long daylengths, but not by intermediate ones.
Apex	Apical part of the shoot including the apical meristem together with the immediate subjacent tissues and young leaf or flower primordia.
Critical daylength	The daylength which separates the inductive and noninductive ranges of photoperiod.
Day-neutral plants	Plants in which flower initiation occurs irrespective of daylength.
Development (of flowers)	Processes occurring between initiation of flowers and anthesis. There is usually no demarcation between initiation and development in most species.
Evocation	Processes in the apex required for irreversible commitment to initiate flower primordia.
Evoked	Adjective given to apices in which evocation is proceeding.
Floral stimulus(i)	Used by us for the factor(s) controlling flower initiation. We refer therefore to chemical and/or physical stimuli (known and hypothetical ones). The floral stimulus as understood in this book may have more than one component and is not necessarily the same in all angiosperms.
Florets	Individual flowers of Compositae and grasses. Also other very small flowers that make-up a very dense form of inflorescence.
Florigen	Name given to the hypothetical hormone that controls flower initiation in all angiosperms.
Induction	Processes required for evocation.
Inductive	Adjective given to treatments that cause induction.
Initiation (of flowers)	Includes the production by meristems of clearly recognizable flower primordia and all preceding reactions that are required if flowers are to be initiated.
Intermediate plants	Plants in which flower initiation is promoted in some intermediate daylength range and inhibited (or delayed) at shorter and longer ones.
Juvenile phase	Period of growth following sowing during which the plants are totally insensitive, or at least poorly sensitive, to conditions which later promote flower initiation.
Juvenility	Character of plants which are in the juvenile phase.
Long-day plants	Plants in which flower initiation is promoted by increasing daylengths. In some species there is a particular value, the critical daylength, above which there is a great increase in flowering.
Long-short day plants	Plants in which flower initiation is promoted by exposure first to LD conditions and then to SD conditions.
Meristem	That part of the shoot apex lying distal to the youngest leaf or flower primordium.
Monocarpic plants	Plants which flower only once in their lifetime.
Photoinduction	Induction by an appropriate photoperiodic regime.
Plastochron	Time interval between the initiation of successive leaf primordia.
Primordium	Refers to a nascent organ with height above its axil.

Short-day plants	Plants in which flower initiation is promoted by reducing daylengths. In some species there is a particular value, the critical daylength, below which there is a great increase in flowering. At very short daylengths, below 6 hr of light per day, initiation may be delayed and this is largely attributable to insufficient photosynthesis.
Short-long-day plants	Plants in which flower initiation is promoted by exposure first to SD conditions and then to LD conditions.
Thermoinduction	Induction by an appropriate low temperature treatment.

ABBREVIATIONS

ABA	(±)-Abscisic acid
ADP	Adenosine diphosphate
AMO-1618	2′-Isopropyl-4′-(trimethylammonium chloride)-5′-methylphenyl-piperidine-1-carboxylate
AMP	Adenosine monophosphate
Ancymidol	α-Cyclopropyl-α-(*p*-methoxyphenyl)-5-pyrimidine methyl alcohol
ATP	Adenosine triphosphate
BA	6-Benzylaminopurine
B-nine	N-(Dimethylamino)succinamic acid, also termed B995, Alar, SADH
c	Cycle(s)
cAMP	Adenosine 3′:5′ cyclic monophosphate (cyclic-AMP)
CCC	2-Chloroethyltrimethylammonium chloride (chlorocholine chloride)
cGMP	Guanosine 3′:5′ cyclic monophosphate (cyclic-GMP)
cpm	Counts per minute
CRP	Cold-requiring plant
cv., cvs.	Cultivar, cultivars
DCA	2,4-Dichloranisole
DCMU	3(3,4-dichlorophenyl)-1,1-dimethylurea
DNA	Deoxyribonucleic acid
DNP	Day-neutral plant
EDDHA	Ethylenediamine-di-*o*-hydroxyphenylacetic acid
EDTA	Ethylenediaminetetraacetic acid
eq	Equivalent
Fe-EDDHA	Iron salt of EDDHA
FR	Far-red light (700 to 770 nm approximately)
GA(s)	Stands for gibberellin(s) in general. Individual gibberellins are designated as GA_1, GA_2, etc.
GA-like substances	Substances occurring naturally and causing responses typical of known GAs, but not yet known chemically
GA_3	Gibberellic acid
GC	Gas chromatography
GC-MS	Combined gas chromatography-mass spectrometry
GLC	Gas-liquid chromatography
GMP	Guanosine monophosphate
hr	Hour(s)
IAA	Indol-3-ylacetic acid
$J.m^{-2}$	Joule per square meter, radiant exposure unit
LD	Long day(s)
LDP	Long-day plant
LSDP	Long-short-day plant
lx	Lux, illuminance unit
min	Minute(s)
NAA	Naphth-1-ylacetic acid, α-naphthalene acetic acid
NADH	Reduced nicotinamide-adenine nucleotide
NADPH	Reduced nicotinamide-adenine dinucleotide phosphate
P	Phytochrome (total)
Pfr	Phytochrome in the far-red absorbing form
Pr	Phytochrome in the red absorbing form

PAR	Photosynthetically active radiation as measured by the photon flux density between 400 and 700 nm
Phosfon(Ph)	2,4-Dichlorobenzyl-tributylphosphonium chloride
R	Red light (600 to 700 nm approximately)
RNA	Ribonucleic acid
mRNA	Messenger RNA
tRNA	Transfer RNA
SD	Short day(s)
SDP	Short-day plant
SK&F-7997	Tris-(2-diethylaminoethyl)phosphate trihydrochloride
SLDP	Short-long-day plant
TIBA	2,3,5-Triiodobenzoic acid
TLC	Thin-layer chromatography
$W \cdot m^{-2}$	Watt per square meter, irradiance unit
wt	Weight

Indexes

SPECIES INDEX

A

B

M

N

O

P

R

S

T

V

W

X

Z

SUBJECT INDEX

A

B

E

F

G

H

I

J

K

L

M

N

O

P

Q

R

U

V

W

X

Y

Z